Magnetoelectric Dipole Antennas

Dedicated to a new class of wideband antenna, significantly developed over the past two decades, this book is the ultimate reference on magnetoelectric dipole antennas. The author is world-renowned for his pioneering work on antennas and has continuously developed the magnetoelectric dipole antenna since 2006. With contributions from the author and his students as well as results from research groups worldwide, the development of this novel antenna is fully captured. The theory and design are presented step-by-step, using simple technical explanations, making the contents accessible to readers without specialized training in antenna designs. Including the various applications of the antenna such as communications, global positioning, sensing, radar, medical imaging, and IoT, this book endeavors to demonstrate the versatility and interdisciplinarity of the antennas.

Helping readers to develop sophisticated antennas with this thorough coverage on magnetoelectric dipole antennas, this is the ideal reference for graduate students, researchers, and electrical engineers.

Kwai Man Luk is Chair Professor in the Department of Electrical Engineering at the City University of Hong Kong. He is a Fellow of the Royal Academy of Engineering and a leading expert in antenna design. His research focuses on microstrip antennas, compact antennas, dielectric antennas, and complementary antenna structures. Over the years, he has received numerous prestigious honors, including the John Kraus Antenna Award in 2017, the Ho Leung Ho Lee Science and Technology Prize in 2019, the Guanghua Engineering Science and Technology Prize in 2022, and the Chen-To Tai Distinguished Educator Award in 2025. He also serves as the principal investigator of a major Area of Excellence project in Hong Kong.

Magnetoelectric Dipole Antennas

KWAI MAN LUK

City University of Hong Kong

CAMBRIDGE
UNIVERSITY PRESS

Shaftesbury Road, Cambridge CB2 8EA, United Kingdom

One Liberty Plaza, 20th Floor, New York, NY 10006, USA

477 Williamstown Road, Port Melbourne, VIC 3207, Australia

314–321, 3rd Floor, Plot 3, Splendor Forum, Jasola District Centre, New Delhi – 110025, India

103 Penang Road, #05–06/07, Visioncrest Commercial, Singapore 238467

Cambridge University Press is part of Cambridge University Press & Assessment,
a department of the University of Cambridge.

We share the University's mission to contribute to society through the pursuit of
education, learning and research at the highest international levels of excellence.

www.cambridge.org
Information on this title: www.cambridge.org/9781108427456

DOI: 10.1017/9781108555715

© Kwai Man Luk 2026

When citing this work, please include a reference to the DOI 10.1017/9781108555715

First published 2026

Cover image: DavidZydd / iStock / Getty Images

A catalogue record for this publication is available from the British Library

A Cataloging-in-Publication data record for this book is available from the Library of Congress

ISBN 978-1-108-42745-6 Hardback

Contents

Preface

Sophisticated and versatile antennas are in huge demand to support the accelerated growth in wireless technologies for communications, remote sensing, radars, health monitoring, and power transfer. The class of microstrip antennas has become the favorite of antenna practitioners, attributing to the attractive features of thin-profile structure, low-cost fabrication, conformability with nonplanar mounting bodies, and compatibility with integrated circuit technologies. The magnetoelectric (ME) dipole can be regarded as an enhanced version of the microstrip antennas. This is due to the fact that while it can be realized as a low-profile printed antenna structure, the ME dipole possesses wider bandwidth, lower cross polarization, lower back radiation, and a more symmetrical radiation pattern, compared with the microstrip antennas. The ME dipole also exhibits stable gain and radiation pattern over the operating frequencies. These attractive features are attributed to the complementary nature of the antenna structure, as an electric dipole mode and a magnetic dipole mode are excited together over the frequency band of operation. It is desirable for graduate students and practicing antenna engineers to gain an in-depth understanding of this new basic antenna element. Although there are a number of book chapters on ME dipoles, there is no single comprehensive reference book available in the market dedicated to this important subject. The proposed book is suitable to be used as a reference book for graduates, teachers, and engineers.

In this book, the design and performance of the original ME dipole antenna is reviewed. The designs of linearly polarized ME dipoles are summarized. Techniques for single-band and dual-band operation are presented. Available feeding structures, including the differential feeds, are studied. Several representatives of dual-polarized and circularly polarized ME dipoles are reviewed. Major techniques for reducing the height and projection area of the ME dipole are examined. Most of the effective designs for ME dipoles with wide bandwidth or ultrawide bandwidth are included. The design principle of each antenna is explained. The development of millimeter-wave 5G and 6G communication systems requires sophisticated high-gain antenna arrays with steerable or reconfigurable radiation patterns. The ME dipole is demonstrated to be an excellent antenna element for realizing these antenna arrays. The designs of millimeter-wave ME dipole arrays with different kinds of feeding structures and transmission lines for diverse applications are reviewed.

Since the introduction of the antenna in 2006, many research groups have contributed to enhance this technology for various applications. The book covers not only the work done by the author and his students but also those useful results achieved by other research groups across the globe. Thus, this book can be regarded as a comprehensive record on the research and development of this new basic antenna. The presented design information is useful for both newcomers and established antenna engineers. I hope that this book will be a welcome addition to the library of any person who is interested in the forefront of antenna technologies.

Through this book, the author wants to convey his deep appreciation to all his students, colleagues, and peers who have helped to advance the ME dipole technology for current and future wireless applications.

1   Introduction

1.1   Background

It is believed that the number of antennas being used in the world is larger than the global population now. This is attributed to the great success in the development of mobile communications over the past two decades. Almost every person has one or more portable phones or devices which have antennas for connectivity with 3G and 4G mobile, Wi-Fi, and GPS wireless networks. In addition, numerous antennas are used in radar and sensor systems. It is expected that the number of antennas will continue to increase exponentially with the development of internet of things and autonomous vehicles.

Stringent requirements have been imposed on the characteristics of antennas used in modern wireless systems. In general, antennas with wide bandwidth and compact size are demanded. For a mobile device, compact miniature wideband antennas that can be operated at multiple bands are required. For a base station, hot spot, or reader, wideband antennas with symmetrical unidirectional radiation pattern with low cross polarization and low back radiation are preferred. The gain and beamwidth of these antennas are required to have small variations over the operating frequencies. Conventional antennas such as the reflector-backed dipole and microstrip patch antennas cannot fulfill these stringent and rigorous requirements.

The magnetoelectric (ME) dipole was invented with the objective to fulfill the requirements imposed by modern wireless systems [1]. It can be considered as a kind of complementary antenna consisting of an electric dipole and a magnetic dipole, which are located closely and perpendicular to each other. At this point, it is worth mentioning that in the limiting case when the dipoles are colocated and very small in size, the complementary antenna is commonly called a Huygens source. In the basic structure of the ME dipole, the electric dipole is a reflector-backed planar dipole and the magnetic dipole is realized by an air-filled quarter-wave shorted-patch antenna. The antenna is excited by an L-shaped probe feed, which is proved to be very effective to increase the bandwidth of different kinds of microwave antennas [2]. It has a size of about 0.6λ (length) $\times$ 0.6λ (width) $\times$ 0.25λ (height), where λ is the resonant wavelength. More than 50% impedance bandwidth for SWR < 2 is achieved. More importantly, the gain of the antenna is over 8 dBi with variation less than 1 dB over the operating frequencies. The E- and H-plane beamwidths are also very stable across the frequency band. All these distinguished features have attracted much research interests in the antenna community since its disclosure in 2006.

A substantial amount of work on the development of ME dipoles has been published by the originator's group and other researchers and scholars over the past decade. It is now the appropriate time to review those findings and put those useful designs into appropriate perspectives. After providing the necessary background in understanding the importance of the ME dipoles in this introductory chapter, the detailed design guideline and performance of various ME dipoles with different characteristics will be presented and discussed in the following chapters.

Microstrip antennas have been developed significantly over the past three decades to support the growth of various wireless applications. They are low in profile which can be made conformable to any mounting body without changing the original shape of the body. This attractive feature enables the microstrip patch antennas to replace other conventional antennas such as the wire monopole, reflector-backed electric dipole, cavity-backed slot antenna, corner reflector antenna, Yagi-Uda antenna, and even dish antenna in many communication and sensor systems. Although the basic microstrip antenna has a length of about half wavelength in dielectric, it is not only useful in base stations but also in small portable devices if dense dielectric material is employed to construct the antenna. In fact, the planar inverted-F antenna that is popularly used in mobile phones and notebooks can be considered as a microstrip antenna with a shorting pin which introduces an additional resonance at a lower frequency.

As the ME dipole is a combination of the reflector-backed electric dipole and the quarter-wave microstrip antenna, a review of the major characteristics of the reflector-backed electric dipole and the microstrip antenna is given in this chapter. The information will provide the readers with a clear picture of how the magnetoelectric dipole was invented to meet the stringent requirement imposed by existing and future wireless systems.

1.2   Reflector-backed Electric Dipole

An electric dipole antenna has a figure-8 shape radiation pattern in the E-plane, which is bidirectional in nature. By adding a rectangular planar reflector of size slightly larger than the electric dipole, a unidirectional antenna can be achieved with high gain if the separation between the electric dipole and the reflector is around a quarter of the operating wavelength. However, the bandwidth of the antenna will be intrinsically narrow. One way to improve the bandwidth of the electric dipole is to increase the width of the two arms of the antenna. A reflector-backed planar dipole operated at around 2.2 GHz with dimensions tabulated in Table 1.1 is depicted in Figure 1.1.

Table 1.1 Dimensions of reflector-backed electric dipole (Unit: mm)

Parameter	L	H	G	W	D	S	C	F
Value/mm	25.5	34	136	15	2	5	1	2

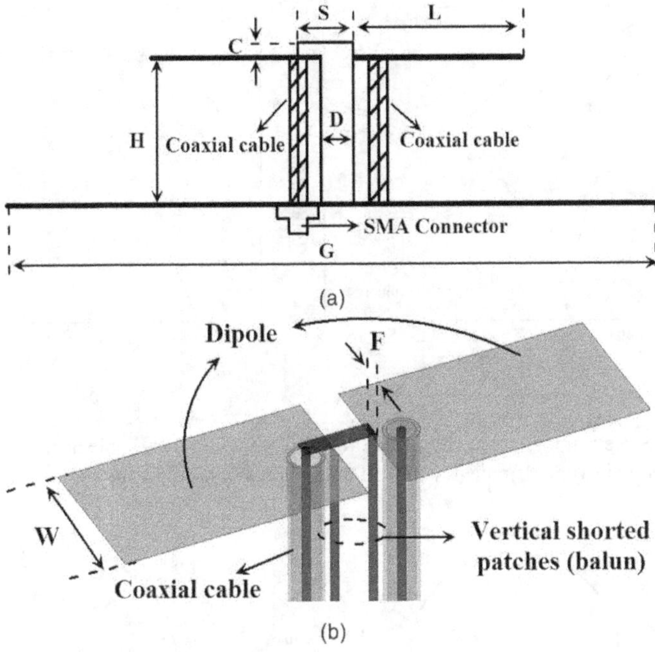

Figure 1.1 Geometry of a reflector-backed electric dipole. (a) Side view and (b) Perspective view

The antenna is purely made of metal. The inner ends of the two arms are connected to the grounded reflector through two vertical narrow strips, which help to form a balun structure. A coaxial cable, located close to one of the narrow strips, is used to transmit the electromagnetic energy from the microwave launcher mounted on the ground plane to excite the inner ends of the two arms. Another identical coaxial cable is added close to the other narrow strip for making the antenna structure more symmetrical for enhancing the symmetry of the co-polarization and reducing the cross-polarization radiation patterns. The width W and the length L of each arm are selected to be about a quarter of the resonant wavelength. The size of the ground plane is about one wavelength square. Other dimensions of the antenna are selected to achieve good impedance matching.

The variation of the performance of the antenna with the width of the two arms is studied. As shown in Figure 1.2, the return loss of the antenna against frequency is presented for the four cases when $b = 5$ cm, 10 cm, 15 cm, and 20 cm. From these curves, the bandwidth of the antenna can be obtained and it is found that the bandwidth is increased from a few percent to over 30% when the width of each arm is increased from 5 to 20 cm. Although this result confirms that the bandwidth of the reflector-backed dipole can be enhanced by increasing the width of the electric dipole, the E-plane and H-plane radiation patterns are not equal and the back-lobe radiation is quite high, as depicted in Figure 1.3. The variation in beamwidth in the E-plane is also substantial, as tabulated in Table 1.2. This finding demonstrates that the wideband reflector-backed dipole has deficiency in performance.

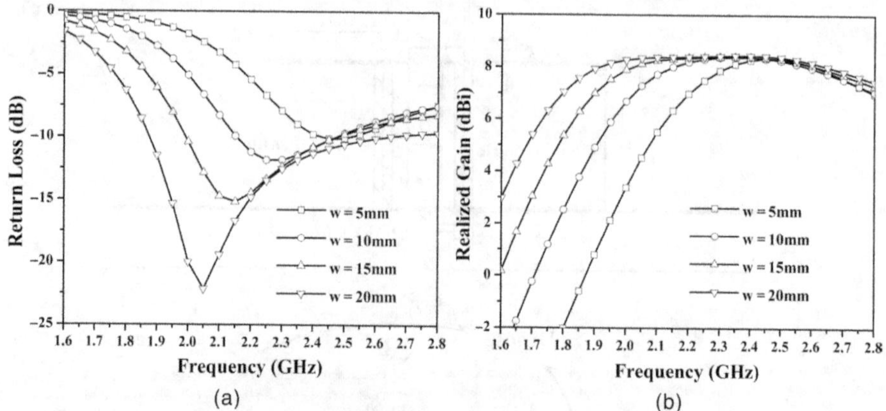

Figure 1.2 (a) Return loss versus frequency for different widths of dipole arms (w), (b) Realized gain versus frequency for different widths of dipole arms (w)

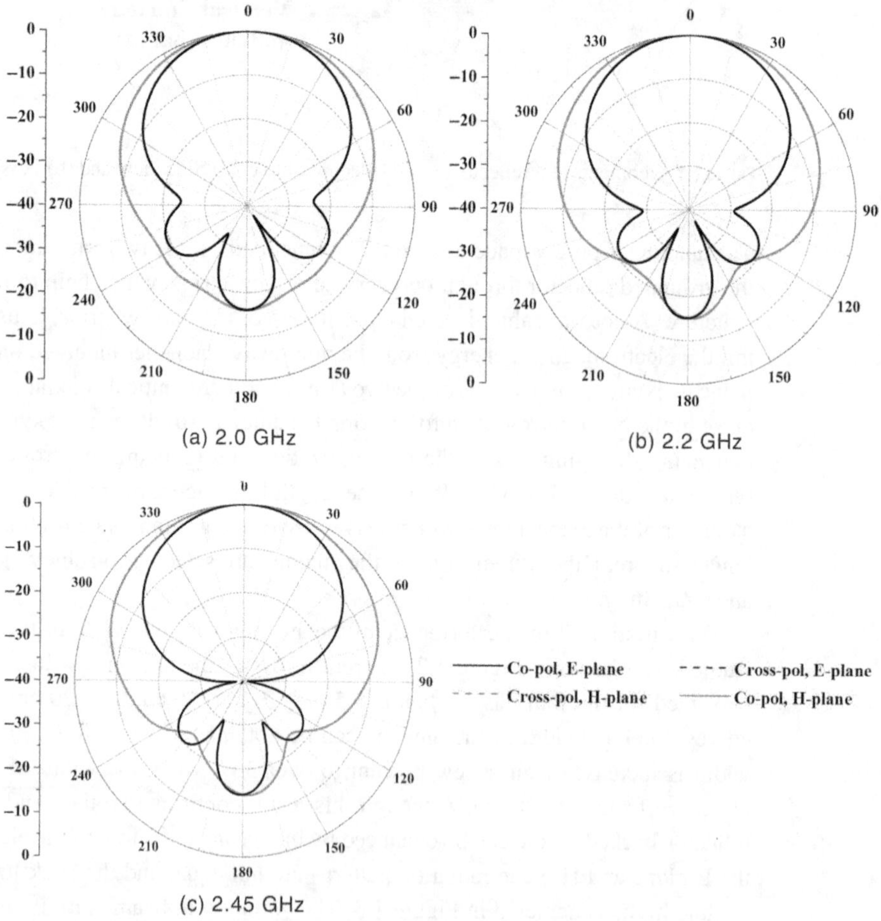

Figure 1.3 Radiation pattern of reflector-backed electric dipole with arm width $b = 15$ mm. (Cross-polarization levels are too small to appear in the figures.) (a) 2.0 GHz, (b) 2.2 GHz, and (c) 2.45 GHz

Table 1.2 3-dB beamwidth in the E- and H-planes

	2.0 GHz	2.2 GHz	2.45 GHz
3-dB beamwidth (E-plane)	61.6°	58.6°	55.6°
3-dB beamwidth (H-plane)	89.8°	88°	90.3°

1.3   Microstrip Antennas

Low-profile microstrip antennas are now popularly used in various wireless systems after three decades of extensive investigation globally since the late 1970s. In its basic form, as shown in Figure 1.4, it consists of a rectangular patch printed on a grounded dielectric substrate. It can be realized using a conventional printed circuit board at low fabrication costs. For operation in the fundamental broadside mode, the length of the antenna, a, is about half a wavelength in dielectric, and the width, b, can be slightly larger than the length to achieve wider bandwidth. No matter what kind of excitation mechanism is employed, such as the probe feed (as depicted in Figure 1.4), microstrip line feed, or aperture coupled feed, the bandwidth is only a few percent of the resonant frequency if the thickness of the dielectric substrate is selected to have a few percent of the resonant wavelength in dielectric. As an example, given the dimensions in Table 1.3, the antenna has about 3% impedance bandwidth (SWR < 2), using the return loss performance as shown in Figure 1.5. This antenna has a peak gain of 7.5 dBi (Figure 1.5) and a unidirectional radiation pattern (Figure 1.6) by simulation.

Many techniques have been proposed to increase the bandwidth of the microstrip antenna in the past. It can be concluded now that those approaches based on either introducing multiple resonances or reducing the Q-factor of the resonance are most effective. Representative examples are the stacked microstrip antenna [3] and the L-probe fed microstrip antenna [4].

By adding a substrate-supported parasitic patch on top of the basic probe-fed single-layer microstrip antenna, a stacked microstrip antenna is obtained as depicted in Figure 1.7. The two rectangular patches are slightly different in sizes for achieving wideband performance. The upper parasitic patch is electromagnetically coupled to the lower patch. This stacking structure can produce double resonances in close proximity. With the dimensions given in Table 1.4, the performance of the antenna is simulated. The antenna has about 7% impedance bandwidth (SWR < 2) and a peak gain of 8 dBi, as shown in Figure 1.8. The radiation pattern does not change significantly as shown in Figure 1.9. Only the back lobe is reduced slightly, which is an advantage.

Another approach to enhance the bandwidth of the microstrip antenna is to increase the thickness of the antenna and to replace the probe feed with a capacitively coupled feed. As shown in Figure 1.10, three dielectric layers are packed together to support a rectangular patch for realizing a thick microstrip antenna, which is excited by an L-shaped probe feed. The horizontal portion of the L-shaped probe – printed on the interface of the two upper dielectric layers – is used to produce a capacitive reactance to counteract the inductive reactance introduced by the vertical portion of the L-shaped probe. In this thick microstrip antenna design, if it is just excited by a

Table 1.3 Dimensions of basic microstrip antenna (Unit: mm)

Parameter	a	b	d	h	g
Value/mm	68	100	50	3.17	136

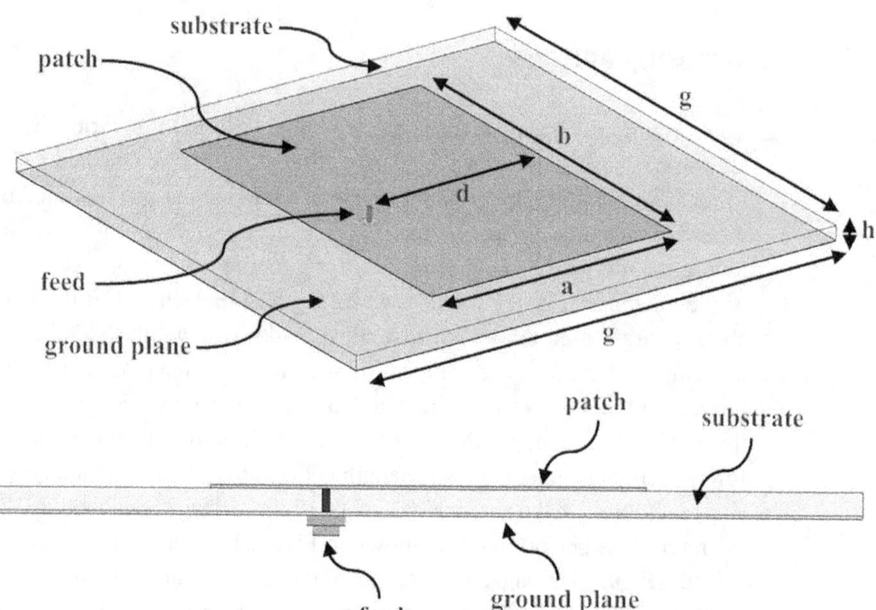

Figure 1.4 Basic microstrip antenna

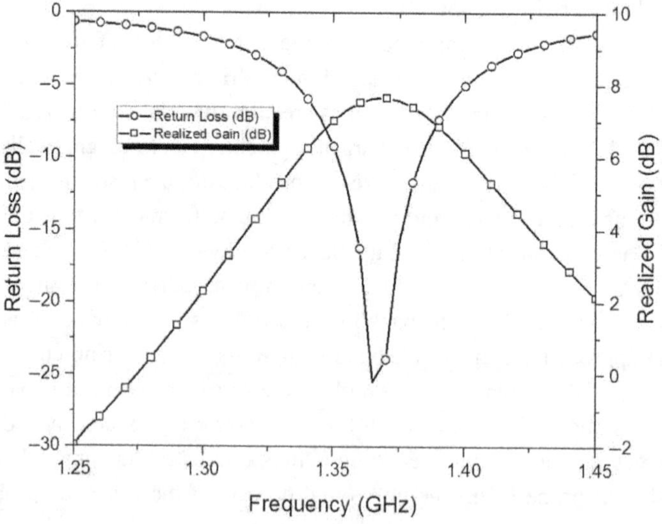

Figure 1.5 Return loss and realized gain of basic microstrip antenna

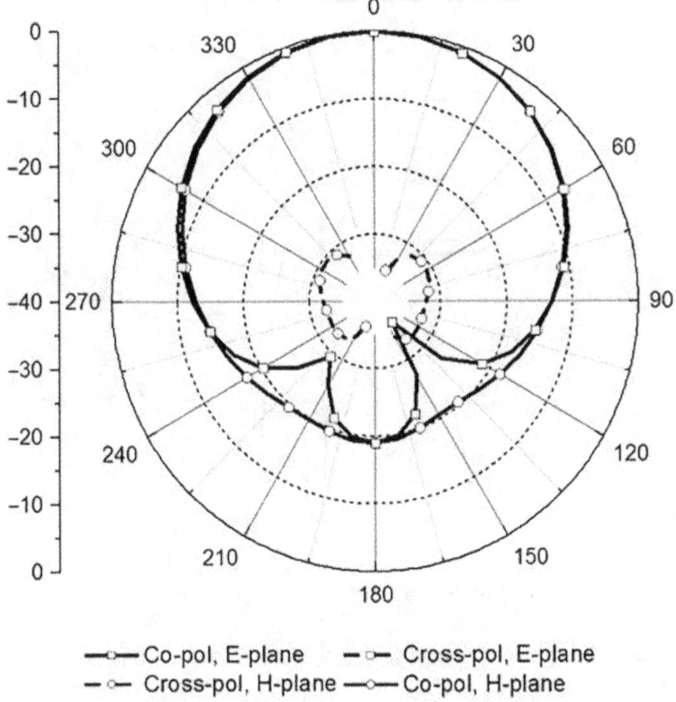

Figure 1.6 Radiation pattern at center frequency of basic microstrip antenna

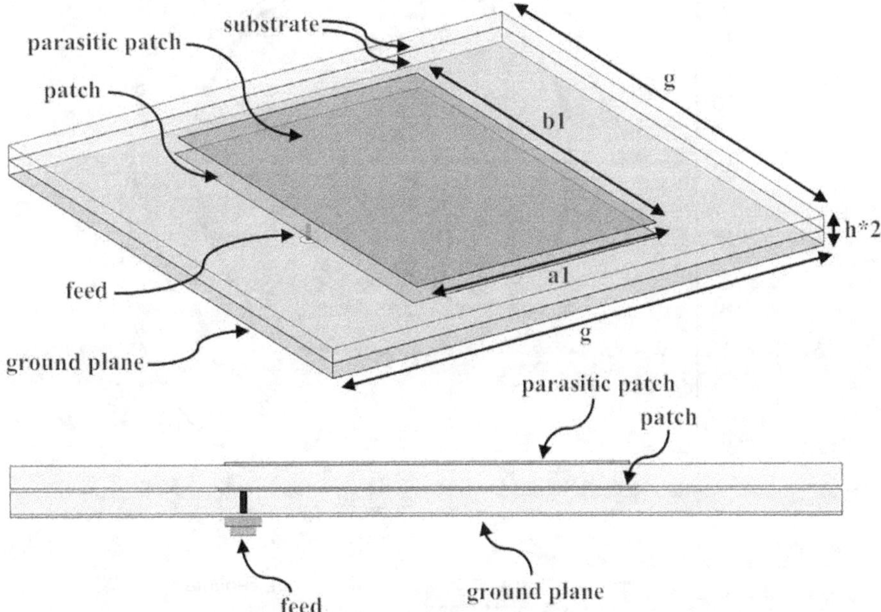

Figure 1.7 Probe-fed stacked microstrip antenna

Table 1.4 Dimensions of stacked microstrip antenna (Unit: mm)

Parameter	a	b	a1	b1	h	g
Value/mm	68	100	66	100	3.17	136

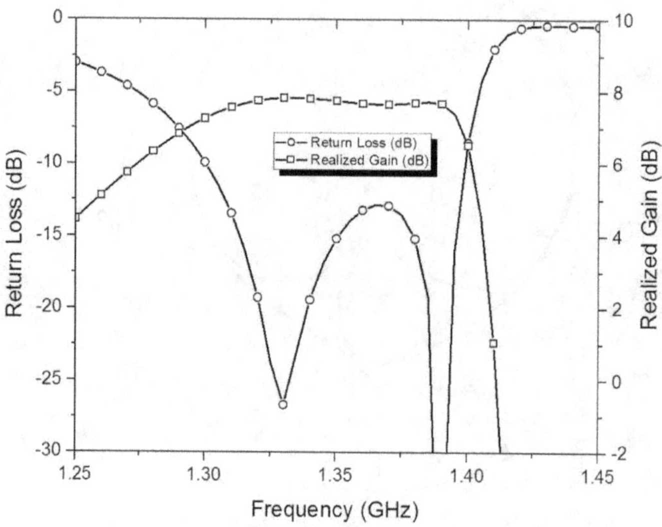

Figure 1.8 Return loss versus frequency of stacked microstrip antenna

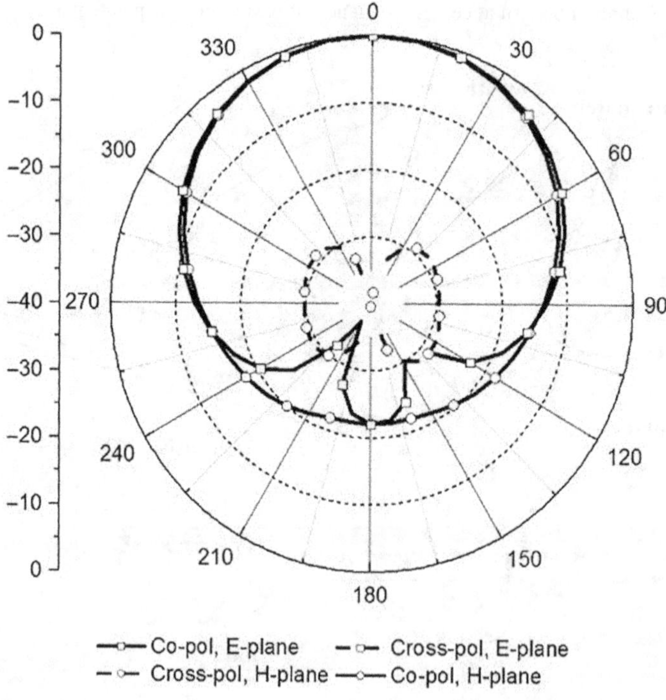

Figure 1.9 Radiation pattern at center frequency of stacked microstrip antenna

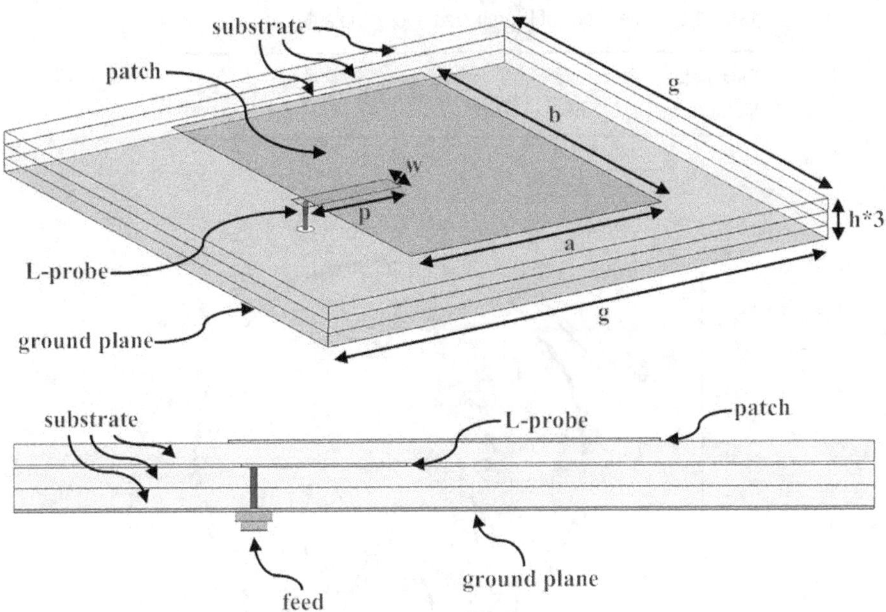

Figure 1.10 Geometry of L-probe fed microstrip antenna

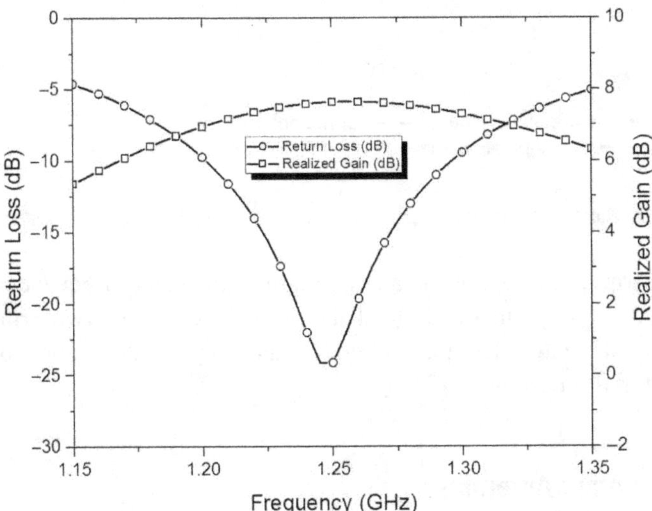

Figure 1.11 Return loss and realized gain of L-probe fed microstrip antenna

vertical coaxial probe, the probe inductance will be high enough to make the antenna completely mismatched.

From the simulated return loss as shown in Figure 1.11, the antenna has an impedance bandwidth of 8% and a peak gain of 7.5 dBi. This demonstrates that the L-probe feed is an effective low-cost approach to enhance the bandwidth of

Table 1.5 Dimensions of L-probe fed microstrip antenna (Unit: mm)

Parameter	a	b	p	w	h	g
Value/mm	68	100	26	6	3.17	136

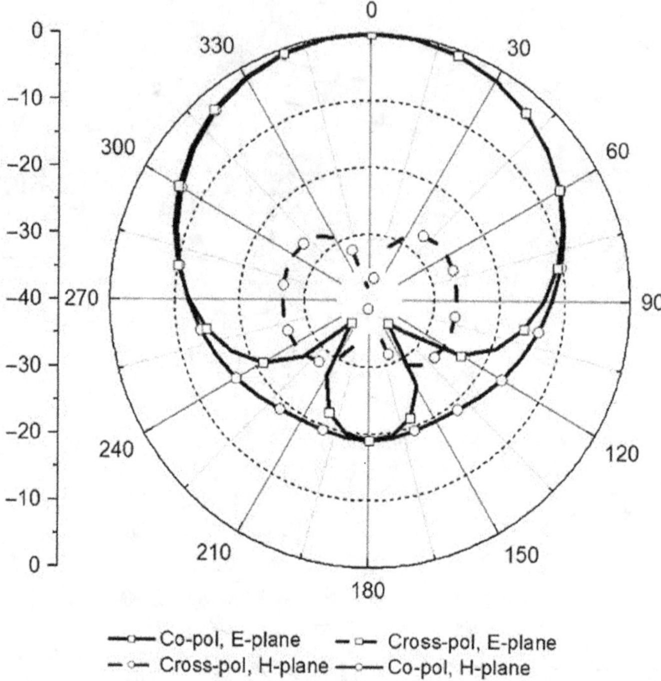

Figure 1.12 Radiation pattern at center frequency of L-probe fed microstrip antenna

the microstrip antenna. The radiation pattern at the center frequency is shown in Figure 1.12. It is symmetrical about the normal axis in the two principal planes with cross polarization less than −25 dB, indicating that the probe radiation is not significant in this design.

1.4   Complementary Antennas

It was shown in 1954 by Clavin [5] that an antenna consisting of two complementary sources of equal radiating power and with proper phase difference can radiate with a radiation pattern that is identical in the E- and H-planes and has no power radiated in the back side direction. This can be proved theoretically for the limiting case by using the far-field formulas of a small electric dipole and a complementary small magnetic dipole [6]. As an illustration, the 3D far-field radiation patterns of a small electric dipole, a small magnetic dipole and the combination of the electric and magnetic

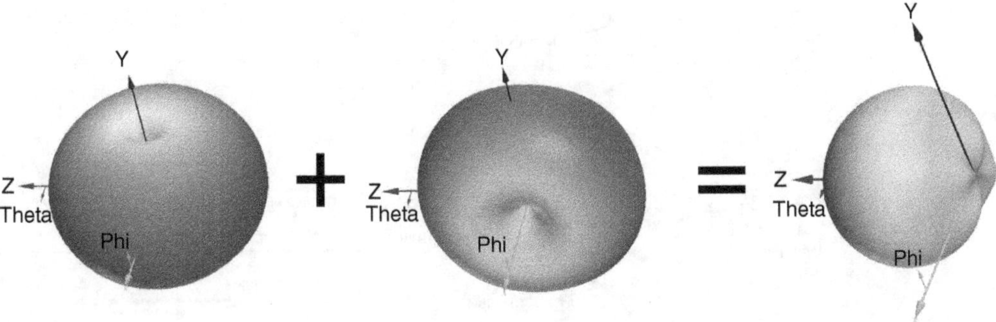

Figure 1.13 Radiation patterns of magnetoelectric dipole

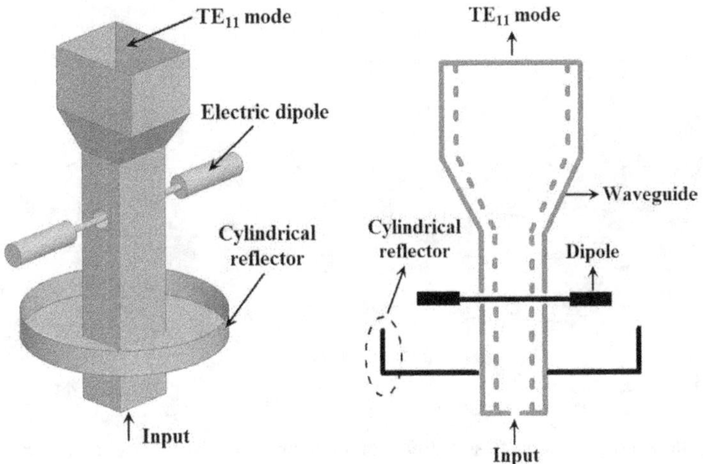

Figure 1.14 The complementary antenna feed proposed by Clavin. Adapted from [5]

dipoles, known as the Huygens source, are depicted in Figure 1.13. It can be clearly seen that the resulting radiation pattern of the complementary antenna looks like a heart shape.

A number of complementary antennas were realized for practical applications. The first one is probably the reflector feed proposed by Clavin [5], as depicted in Figure 1.14. It consists of a cylindrical electric dipole and an open end of a circular waveguide operating in the TE11 mode. In performance, it has equal E- and H-plane patterns over 22% with SWR < 1.5 if proper dimensions for the electric dipole and the TE11 source are selected. The back radiation of this feed is below −30 dB.

In 1960, King and Owyang [7] studied analytically and experimentally the effect of parasitic dipoles on the input impedance and radiation pattern of a narrow slot antenna, as shown in Figure 1.15. It was demonstrated that the radiation pattern shape in both principal planes can be changed substantially by adjusting the separation between the electric dipoles and the slot. The bandwidth of the antenna, however, was not presented.

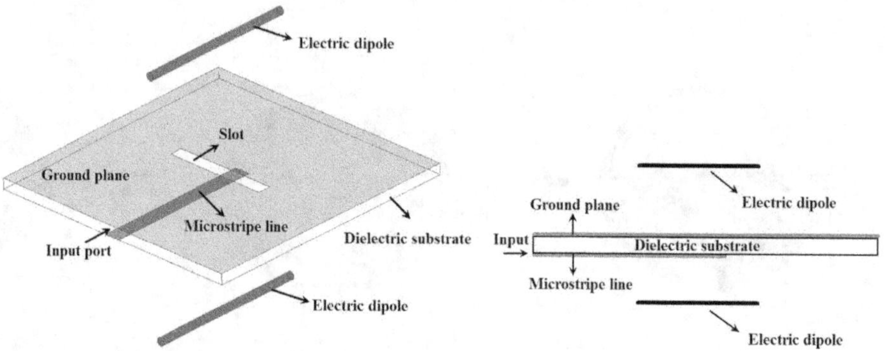

Figure 1.15 A slot antenna with coupled electric dipoles proposed by King and Owyang. Adapted from [7]

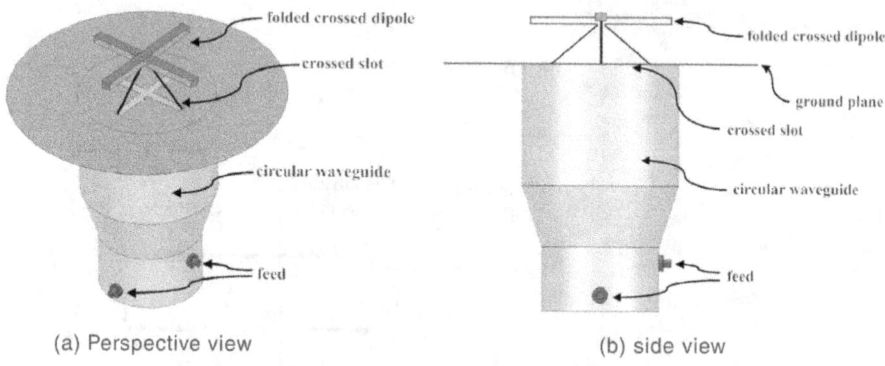

Figure 1.16 A complementary slot-dipole antenna proposed by Gabriel and Dod. (a) Perspective view and (b) Side view. Adapted from [8]

In 1966, an S-band complementary slot-dipole antenna with hemispherical coverage for the Apollo reentry tracking interferometer was reported by Gabriel and Dod from NASA Goddard Space Flight Center [8]. As shown in Figure 1.16, a folded crossed dipole is mounted above a waveguide-excited crossed slot antenna. Both antenna elements are excited independently for ease of adjusting the power division and phase difference between the two elements.

In 1974, Clavin disclosed another complementary antenna element which is suitable for use as an antenna array element [9]. The antenna structure, known as the Clavin element [10], is very simple in construction. As shown in Figure 1.17, a pair of inverted L-shaped wires is connected to the two sides of the ground plane of a narrow slot antenna, which is excited by a rectangular waveguide below the ground plane. Effectively, both the slot and the inverted-L wires are excited by a single feed line. It was demonstrated that in addition to achieving equal E- and H-plane patterns, this complementary antenna has more attractive features. The beamwidths can be varied by changing the lengths of the inverted-L wires. In particular, a $\cos\theta$ pattern can be achieved. It is worth mentioning that an array of elements with $\cos\theta$ pattern has the

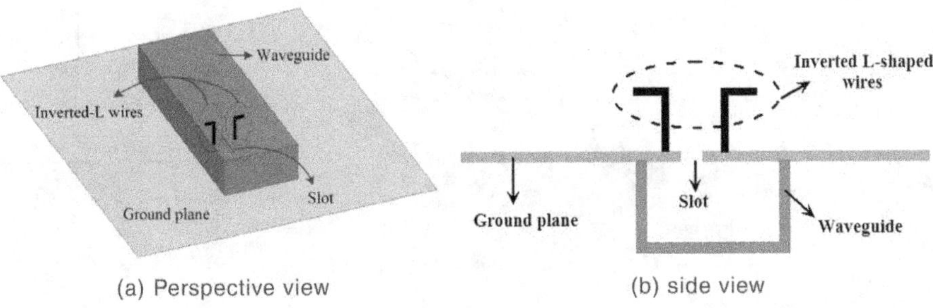

Figure 1.17 The Clavin element. (a) Perspective view and (b) Side view. Adapted from [10]

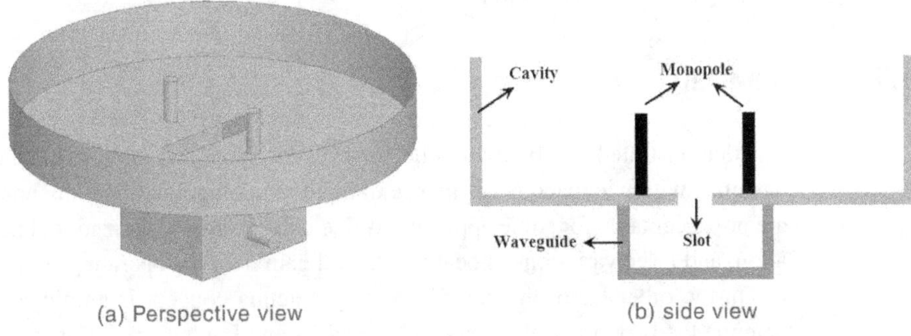

Figure 1.18 A complementary multimode antenna by Clavin. (a) Perspective view and (b) Side view. Adapted from [11]

input resistance invariant with the scan angle of the radiation beam. The nulls in the two principal planes can be placed along the ground plane, which can substantially suppress the mutual impedance between array elements. This antenna has a bandwidth of about 10%.

In 1975, the above antenna was further modified by Clavin to realize an antenna feed with circularly symmetric radiation pattern exhibiting equal pattern shape in both E- and H-planes and very low back radiation [11]. As shown in Figure 1.18, the monopole-loaded slot is placed inside an open-end circular waveguide for generating the complementary TM11 and TE11 modes. Excellent radiation patterns were achieved by optimizing the height and diameter of the cup reflector and the height and separation of the two vertical wires.

In 1972, a broadband unidirectional antenna pointing to the horizontal direction consisting of a vertical monopole and a horizontal cavity-backed slot was revealed by Mayes et al. [12]. As shown in Figure 1.19, the monopole and slot are excited by a straight microstrip line with one end connected to the source and the other end connected to a matched load. With this arrangement, the antenna can achieve over bandwidths of 10 to 1 or more. The antenna, however, has radiation efficiency less than 10% at the lower operating frequencies. The directional pattern is not pointing to the broadside direction, limiting its application areas.

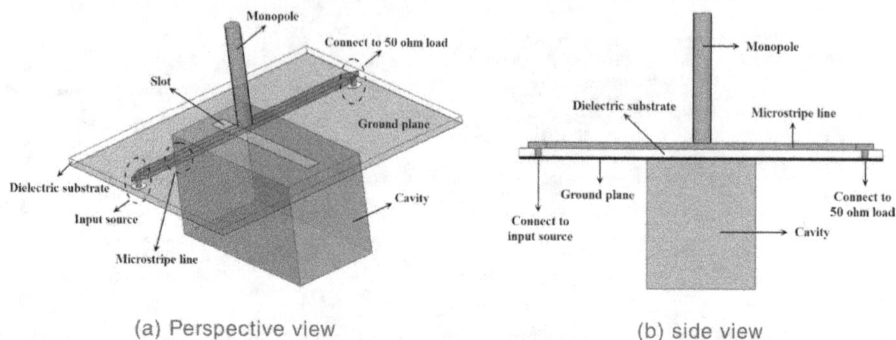

(a) Perspective view (b) side view

Figure 1.19 A broadband monopole slot antenna proposed by Mayes et al. (a) Perspective view and (b) Side view. Adapted from [12]

1.5 Summary

It is demonstrated that both the reflector-backed dipole and microstrip antenna can be designed with wideband performance in input matching, but other antenna parameters are not acceptable for some applications due to high cross polarization, high back radiation, and large variations in beamwidth and gain over the operating frequency band.

On the other hand, the complementary antenna concept is employed to develop antenna feeds or array elements with equal E- and H-plane radiation pattern and low back radiation. All the designs available in the literature exhibit excellent radiation pattern characteristic. However, the bandwidth of most of these antennas is wide enough for modern wireless applications. The design by Mayes et al. is very wide, but it is not a broadside antenna, and the efficiency is very low at the lower operating frequencies.

Most of the classical complementary antennas are narrow in bandwidth. This is probably due to the fact that narrow slots were used to realize magnetic currents. The ME dipole is proposed based on the complementary antenna concept with the objective to develop wideband unidirectional antennas with broadside radiation. This is achievable as the magnetic current is realized by a wideband microstrip patch antenna, together with the use of a planar electric dipole with a large width for wideband performance.

References

[1] K. M. Luk and H. Wong, A new wideband unidirectional antenna element, *International Journal of Microwaves and Optical Technologies*, vol. 1, 2006, no. 1, pp. 35–44.

[2] N. Nakano, M. Yamazaki and J. Yamauchi, Electromagnetically coupled curl antenna, *Electronics Letters*, vol. 33, 1997, pp. 1003–1004.

[3] S. A. Long and M. Walton, A dual-frequency stacked circular-disc antenna, *IEEE Transactions on Antennas and Propagation*, vol. 27, 1979, no. 2, pp. 270–273.

[4] K. M. Luk, C. L. Mak, Y. L. Chow and K. F. Lee, Broadband microstrip patch antenna, *Electronics Letters*, vol. 34, 1998, no. 15, pp. 1442–1443.

[5] A. Clavin, A new antenna feed having equal E- and H-plane patterns, *IRE Transactions – Antennas and Propagation*, vol. 2, 1954, pp. 113–119.

[6] M. Li and K. M. Luk, Wideband magnetoelectric dipole antennas, *Handbook of Antenna Technologies*, Springer Science+Business Media Singapore 2016, pp. 1969–2017.

[7] R. W. P. King and G. H. Owyang, The slot antenna with coupled dipoles, *IRE Transactions on Antennas and Propagation*, vol. 8, 1960, pp. 136–143.

[8] W. F. Gabriel and L. R. Dod, A Complementary Slot-Dipole Antenna for Hemispherical Coverage, Report of NASA-Goddard Space Flight Center, Greenbelt, Maryland, NASA TM X-55681, Oct. 1966.

[9] A. Clavin, D. A. Huebner and F. J. Kilburg, An improved element for use in array antennas, *IEEE Transactions on Antennas and Propagation*, vol. 22, 1974, no. 4, pp. 521–525.

[10] R. S. Elliott, On the mutual admittance between Clavin elements, *IEEE Transactions on Antennas and Propagation*, vol. 28, 1980, no. 6, pp. 864–870.

[11] A. Clavin, A multimode antenna having equal E- and H-Planes, *IEEE Transactions on Antennas and Propagation*, vol. 23, 1975, no. 5, pp. 735–737.

[12] P. E. Mayes, W. T. Warren and F. M. Wiesenmeyer, The monopole-slot: A small broadband, unidirectional antenna, *IEEE Transactions on Antennas and Propagation*, vol. 20, 1972, no. 4, pp. 489–493.

2 The Basic Magnetoelectric Dipole

2.1 Introduction

The magnetoelectric dipole, a name suggested by the late Prof. K. K. Mei, can be considered as a kind of complementary antenna which is composed of colocated electric and magnetic dipoles. It is different from the self-complementary antenna which has a more stringent requirement on the antenna structure. The Huygens source, consisting of a point electric dipole and a point magnetic dipole, is also a complementary antenna in the limiting case. The classical complementary antenna proposed by A. D. Clavin [1] in the early 1970s consists of a waveguide-fed slot antenna with a parasitic dipole to achieve symmetrical radiation patterns with low cross polarization. The antenna is narrow in bandwidth and limited in applications as it is fed by a bulky waveguide, which is not convenient in low-frequency applications.

By combining a wideband quarter-wave patch antenna with a planar reflector-backed electric dipole together, the wideband magnetoelectric dipole was proposed in 2006. The antenna has an impedance bandwidth (SWR < 2) close to 50%, which is wider than that of the patch and dipole antennas. More importantly, this linearly polarized magnetoelectric dipole has a stable radiation pattern and gain over the operating frequencies, with the co-polarized radiation pattern highly symmetrical and very low cross polarization.

In this chapter, the performance of the basic linearly polarized magnetoelectric dipole is reviewed in detail to prepare the readers to appreciate other sophisticated designs in the chapters to follow. A new equivalent circuit of the antenna is given, which is different from the previous one proposed in the literature [2]. The current density distributions on the antenna surfaces are provided to help understand the operating principle of the magnetoelectric antenna. The effect of ground plane size and side wall height on the radiation patterns is given. Finally, a design guideline is suggested.

2.2 Basic Structure

The structure of the basic magnetoelectric dipole is shown in Figure 2.1. It consists of a horizontal planar half-wave electric dipole supported by two vertical quarter-wave electric walls which are mounted on a horizontal ground plane. The two electric walls, together with the portion of ground plane connecting them, can operate like a shorted

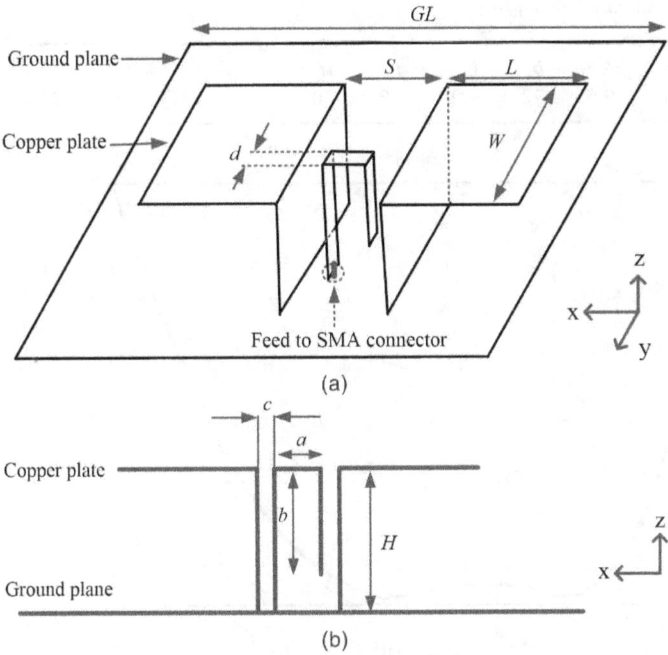

Figure 2.1 Geometry of the basic magnetoelectric dipole. (a) Perspective view and (b) Side view. Adapted from [4]

patch antenna. For wideband operation, the separation between the two vertical walls is chosen as 14% wavelength. The radiation of the shorted patch antenna is mainly from the horizontal open end which radiates as a magnetic current. A Γ-shaped strip is selected as the feed of the antenna. The first portion of the feed serves as a 50-ohm transmission line with one end connected to the coaxial launcher mounted below the ground plane and the other end connected to the second portion of the feed which is responsible for exciting the electric and magnetic dipoles. Since the second portion has inductive effect which affects the matching performance of the antenna, the third portion of the feed that works as an open-end transmission line is added to introduce a capacitance to compensate the effect due to the feed inductance. The second and third portions of the strip together serve as the conventional L-probe feed proposed by Nakano [3].

2.3   Performance of the Antenna

The size of the antenna, as tabulated in Table 2.1, was chosen after carrying out an optimization process using a commercial solver, which has a parameter slightly different from that presented in the original paper published in 2006 [4]. The difference is that the height of the 50-ohm line (portion 1) is selected as $c = 1.2$ mm instead of $c = 1$ mm. It can be observed that the SWR is less than 1.5 over a wide frequency range from 1.8 to 2.75 GHz, equivalent to 42% bandwidth, as shown in Figure 2.2.

Table 2.1 Optimized dimensions

Parameters	A	b	C	d	H	L	S	W	GL
Value/mm	9.5	22	1.2	4.9	30	30	17	60	120

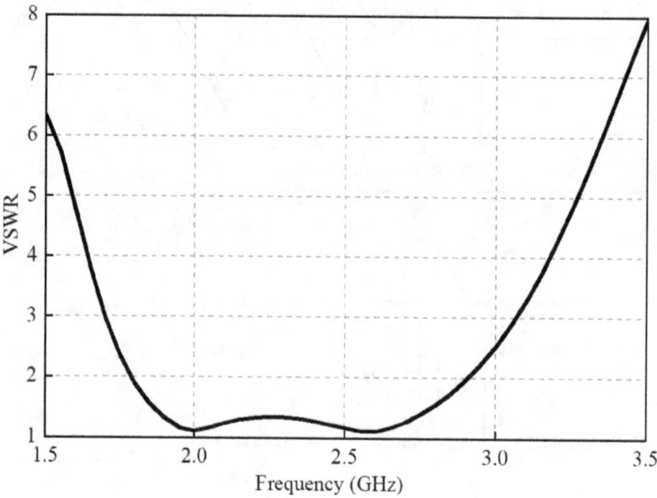

Figure 2.2 SWR versus frequency

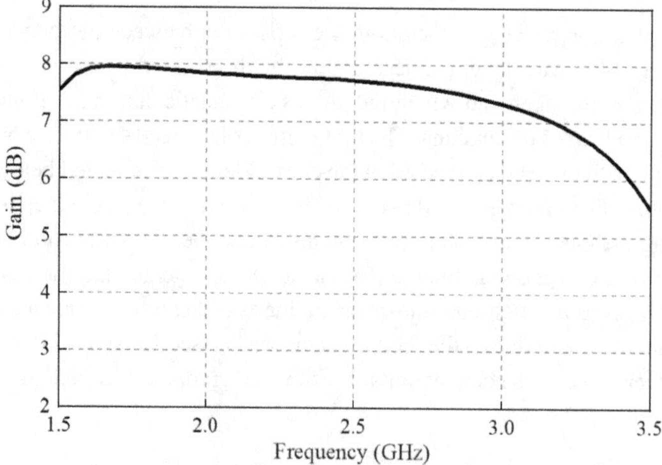

Figure 2.3 Gain versus frequency

The gain performance is depicted in Figure 2.3. It can be observed that the gain varies from 7.6 to 8 dBi over the operating frequency range from 1.8 to 2.75 GHz. This is one of the attractive feature of the magnetoelectric dipole, as the gain variation is just ±0.4 dB over 42% bandwidth. The radiation patterns of the antenna at different frequencies are depicted in Figure 2.4. It can be observed that the patterns

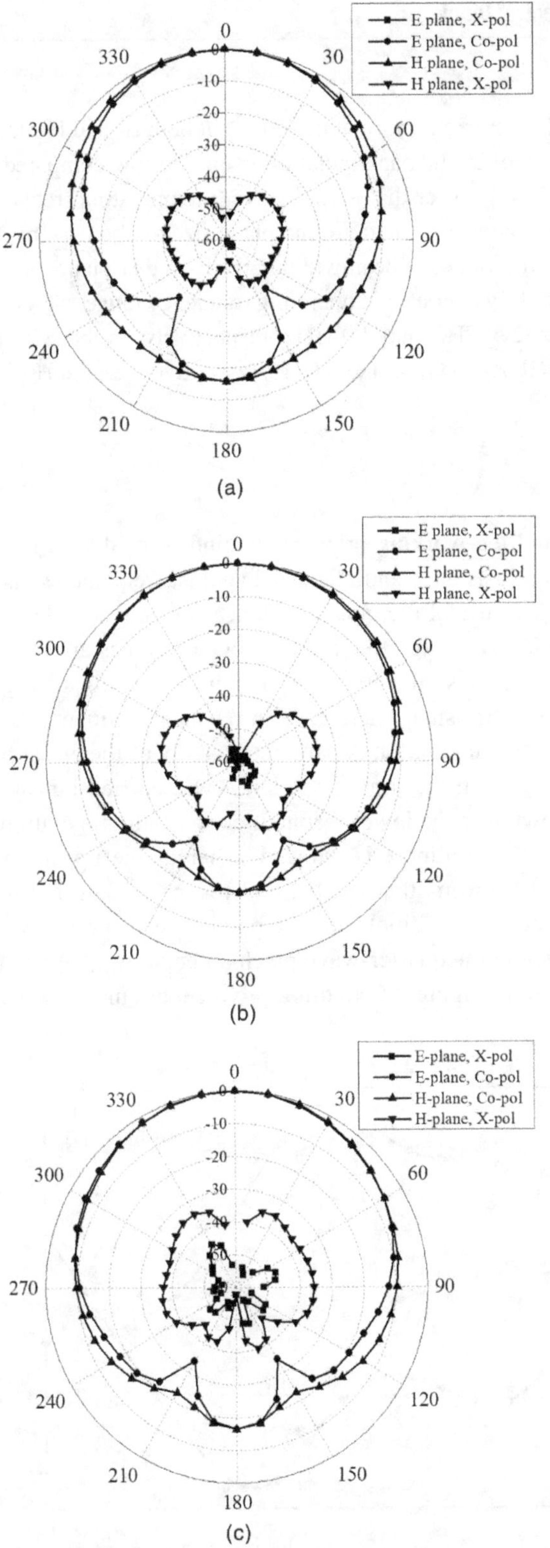

Figure 2.4 Radiation patterns of the magnetoelectric dipole at different frequencies. (a) 1.75 GHz, (b) 2.5 GHz, and (c) 3.0 GHz

are symmetrical to the boresight (normal direction) in both E- and H-planes with low cross polarization. Although the back radiation is not too low compared to the theoretical prediction, it is still lower than −18 dB over the operating frequencies, which is a significant achievement as a small ground plane size of about 1λ by 1λ is used.

More importantly, the values of beamwidth in the E- and H-planes are almost identical and vary only slightly over a wide operating frequency range, which are 70°, 75°, and 75° at 1.75 GHz, 2.5 GHz, and 3.0 GHz, respectively, in the E-plane, and 72°, 75°, and 75° at 1.75 GHz, 2.5 GHz, and 3 GHz, respectively, in the H-plane.

2.4   Discussions

In order to understand the working principle behind, it is always better to check the variations of input resistance and reactance against frequencies, as depicted in Figure 2.5. Two peak values of the input resistance can be observed at frequencies of 2.00 GHz and 2.60 GHz, where the input reactance values are very low in values. This behavior can be described by an equivalent circuit consisting of two parallel R-L-C resonators with small difference in resonant frequencies, and the two resonators are connected in parallel. A parametric study of the effect of the length of the electric dipole, L, on the input impedance was carried out in the original paper [4]. It was shown that the lower resonant peak would move up in frequency if the length of the electric dipole is reduced, whereas the upper resonant peak was unchanged. This confirms that the parallel resonator with a lower resonant frequency is due to the electric dipole and the parallel resonator with a higher resonant frequency is due to the quarter-wave patch antenna. An equivalent circuit of the antenna is depicted in Figure 2.6. In this equivalent circuit, the series resonator

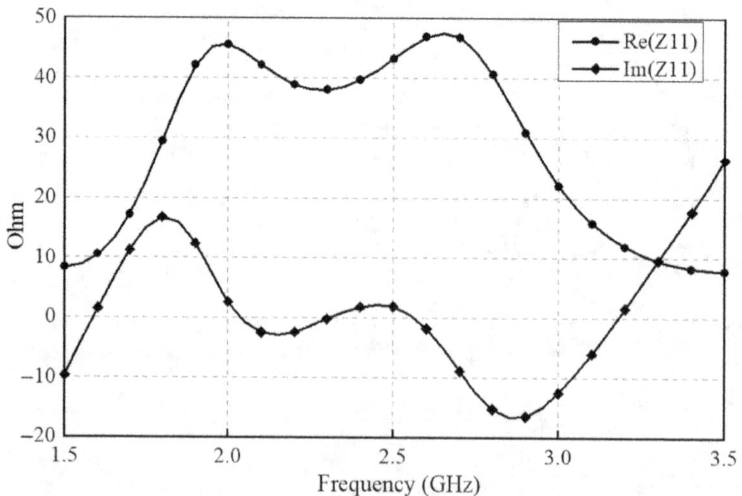

Figure 2.5 Input impedance against frequency

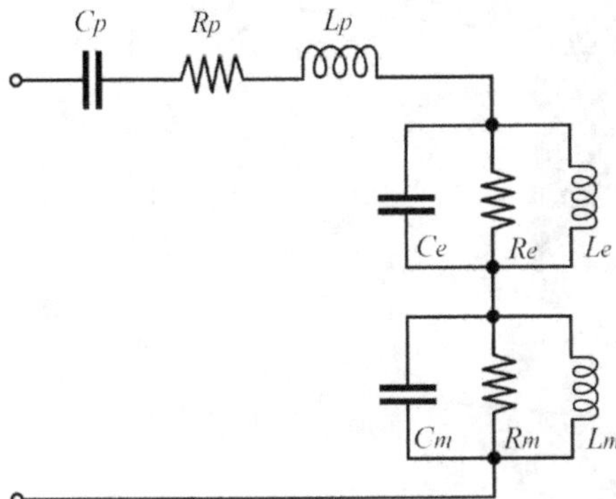

Figure 2.6 Equivalent circuit of the magnetoelectric dipole

R_p–L_p–C_p represents the L-shaped probe feed, the parallel resonator, R_e–L_e–C_e, represents the electric dipole, and the parallel resonator, R_m–L_m–C_m, represents the quarter-wave patch antenna.

To further understand the operation of the magnetoelectric dipole, the current density distributions on the metallic surfaces of the antenna at different times and different frequencies are obtained as shown in Figure 2.7. At a reference time $t = 0$, the current in the inner surfaces of the vertical walls is strongly excited at $f = 2.05$ GHz, the first resonant point (Figure 2.7(a)). It has a quarter-wavelength distribution with the maximum and minimum values occurred at the bottom and the top of each vertical wall, respectively. The current is very weak on the two horizontal plates at this time. At $t = T/4$, where T is the period of time, the current in the two vertical walls decreases to zero except over a small region under the feed line, whereas the current in the two horizontal plates is strongly excited. The variation of the current density in the two horizontal plates as a whole follows a half-wavelength distribution with the maximum occurring at the inner edges (Figure 2.7(b)). The same behavior of current density variation is found at the second resonant point at $f = 2.55$ GHz (Figure 2.7(c) and (d)). The result indicates that both the electric dipole and quarter-wave patch antenna are excited over a wide operating frequency range. The current densities for the electric dipole and quarter-wave patch antenna have a 90° phase difference.

2.5　Effect of Ground Plane Size and Structure

Theoretically, the magnetoelectric dipole should have very low back radiation. The original magnetoelectric dipole, however, is not too low in backlobe level. So, the effect of the ground plane size and adding side walls on the radiation patterns of the magnetoelectric dipole was investigated. It can be observed in Figure 2.8(b), (c), and (d) that the backlobe is increased by about 3 dB at the center and upper frequency points when the length of the ground plane is decreased from 120 to 100 mm. The increase in backlobe level is more substantial at the lower frequency point. It was

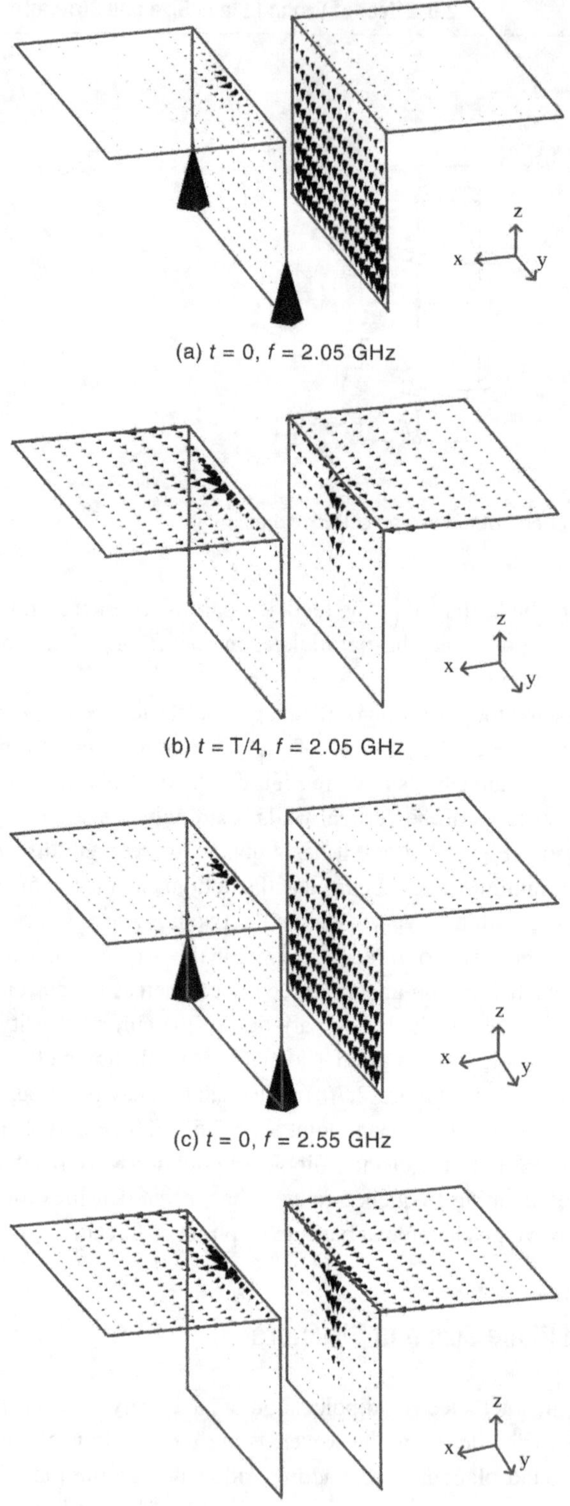

Figure 2.7 Current density distributions on antenna surfaces. (a) $t = 0$, $f = 2.05$ GHz, (b) $t = T/4$, $f = 2.05$ GHz, (c) $t = 0$, $f = 2.55$ GHz, and (d) $t = T/4$, $f = 2.55$ GHz

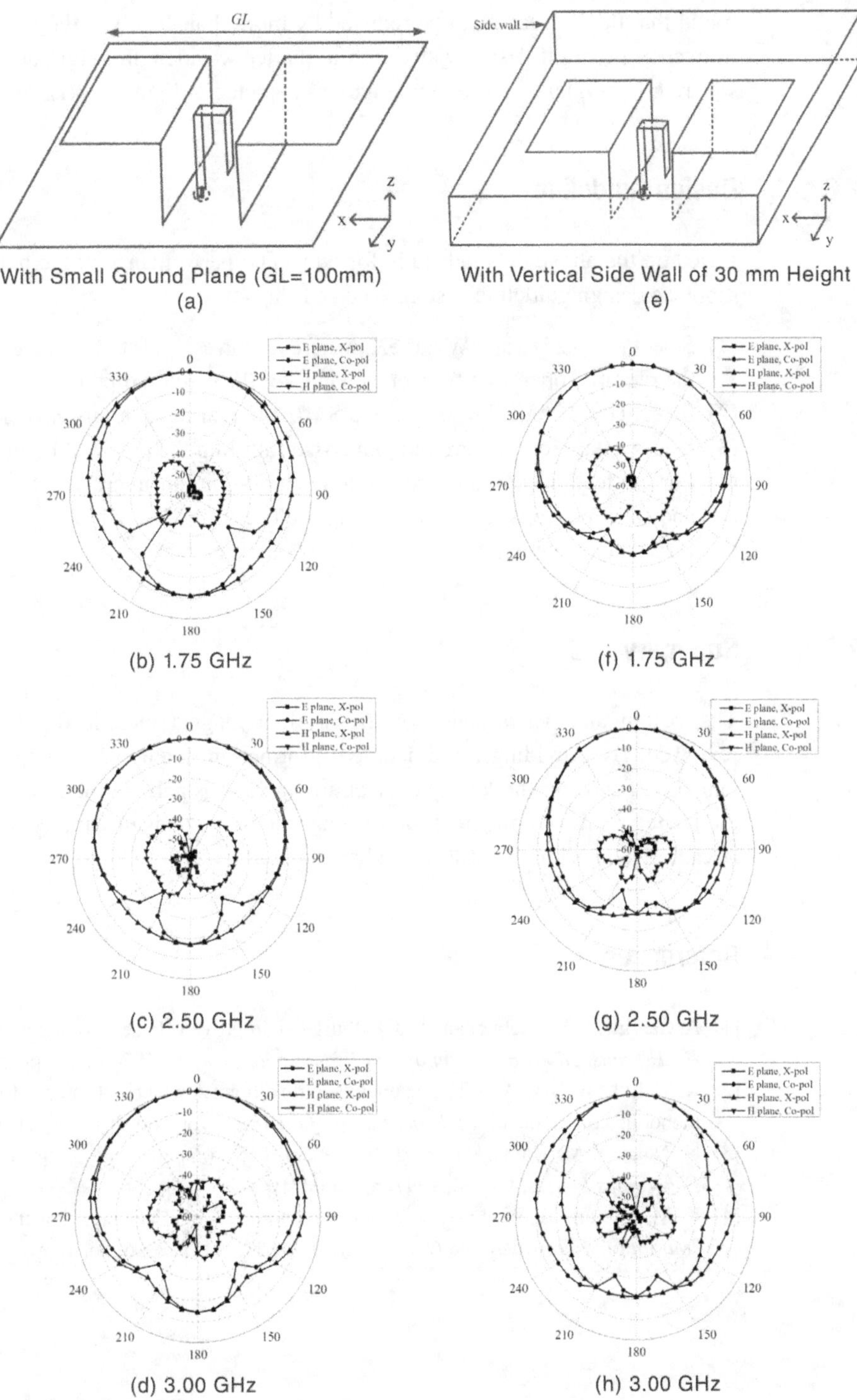

Figure 2.8 Radiation pattern of the magnetoelectric dipole with different ground plane size and structure. (a) With small ground plane (GL = 100 mm), (b) 1.75 GHz, (c) 2.50 GHz, (d) 3.00 GHz, (e) with vertical side wall of 30 mm height, (f) 1.75 GHz, (g) 2.50 GHz, and (h) 3.00 GHz

found that the backlobe can be reduced by more than 10 dB at the lower and center frequency points if 4 side walls with a quarter-wavelength height are added to the square ground plane of 120 mm length, as depicted in Figure 2.8(f), (g), and (h).

2.6   Design Guideline

Based on the above detailed study for wideband performance of the magnetoelectric dipole, a design guideline is suggested as follows:

(1) Select $L = 0.25\lambda_0$ and $W = 0.5\lambda_0$, where λ_0 corresponds to the center frequency of the required operating frequency band, for the electric dipole.
(2) Select $H = 0.25\lambda_0$, $W = 0.5\lambda_0$, and $S = 0.14\lambda_0$ for the quarter-wave patch antenna.
(3) Select an appropriate ground plane size fulfilling the practical requirement.
(4) For the feed design, the first portion is a 50-ohm air microstrip line. The second and third portions have a total length close to $0.25\lambda_0$, which form an L-shaped feed that can be adjusted for good impedance matching.

2.7   Summary

The design and performance of the original magnetoelectric dipole antenna are reviewed. By providing more detailed information, including an improved equivalent circuit and current density distribution in various parts of the antenna, hopefully, the readers will have the necessary background knowledge to appreciate those more advanced designs in the following chapters.

References

[1] A. Clavin, D. A. Huebner and F. J. Kilburg, An improved element for use in array antennas, *IEEE Transactions on Antennas and Propagation*, vol. 22, 1974, no. 4, pp. 521–525.
[2] K. M. Luk and B. Q. Wu, The magnetoelectric dipole: A wideband antenna for base stations in mobile communications, *Proceedings of the IEEE*, vol. 100, 2012, no. 7, pp. 2297–2307.
[3] N. Nakano, M. Yamazaki and J. Yamauchi, Electromagnetically coupled curl antenna, *Electronics Letters*, vol. 33, 1997, pp. 1003–1004.
[4] K. M. Luk and H. Wong, A new wideband unidirectional antenna element, *International Journal of Microwaves and Optical Technologies*, vol. 1, 2006, no. 1, pp. 35–44.

3 Magnetoelectric Dipoles with Linear Polarization

3.1 Introduction

In this chapter, the development of linearly polarized magnetoelectric (ME) dipoles operated at lower microwave frequencies is reviewed. Magnetoelectric dipoles can be fabricated at low costs as they are purely made of metal plates in the few GHz range. Designs with modified L-shaped probe feeds for various purposes are first presented. Magnetoelectric dipoles with modified dipole shapes and feeds for enabling the antennas to be DC grounded are summarized. The aperture coupling technique was widely applied for the designs of microstrip antennas. Magnetoelectric dipoles with aperture-coupled feeds were also proposed in the literature. Their characteristics are presented. Differentially fed ME dipoles are also reviewed. The performance of ME dipoles for MIMO systems is discussed, which is of topical interest for 5G applications. Some recent applications of linearly polarized ME dipoles in different array environments are also presented.

3.2 ME Dipoles with Modified L-Probe Feeds

For ease of fabrication of the ME dipole, the L-shaped probe can be replaced by a T-shape probe, as shown in Figure 3.1. The width of the antenna is kept at $W = 0.5\lambda_0$. The lengths of the two branches of the T-shaped probe produce capacitive effect to counteract the inductive effect due to the central section of the probe. It is found that the bandwidth of the ME dipole can be increased from about 53% to 66% when the L-probe feed is replaced by a T-probe feed, with other characteristics almost unchanged [1]. The gain is also kept at around 8 dBi.

The ME dipole can be excited by two parallel L-shaped feeds as shown in Figure 3.2. Using this twin L-probe technique that was used in the design of wideband microstrip antennas, the ME dipole has slightly improved performance of 70% impedance bandwidth and 8.5 dBi antenna gain. The variation of beamwidth over the operating frequencies in the E-plane and H-plane is, respectively, about 6° and 15° [1].

For achieving a simplified feed structure, the antenna configuration can be modified as shown in Figure 3.3. The antenna can be considered as a combination of two parallel electric dipoles and one magnetic dipole. The antenna preserves all good

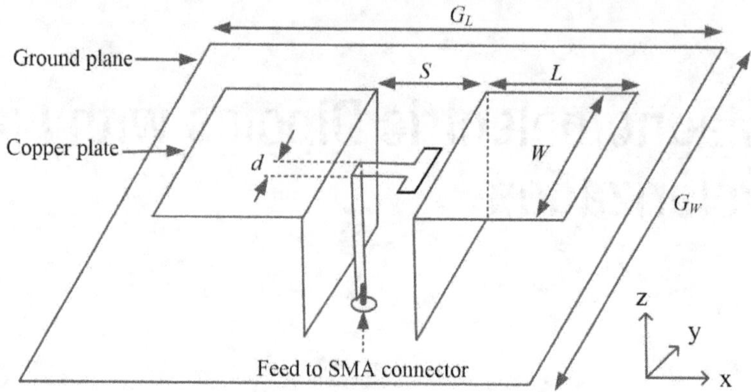

Figure 3.1 Magnetoelectric dipole with T-shaped feed

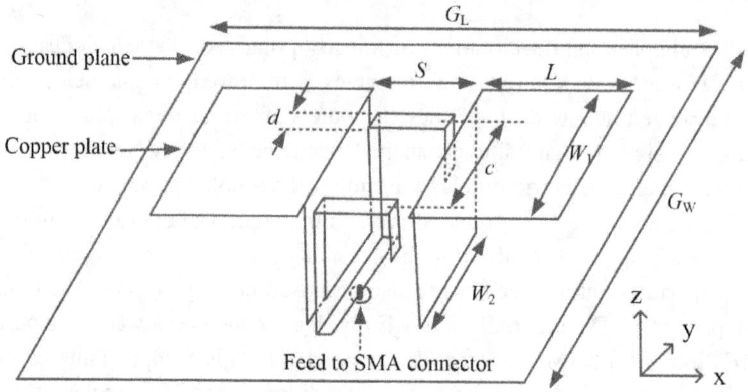

Figure 3.2 Magnetoelectric dipole with twin L-shaped feeds

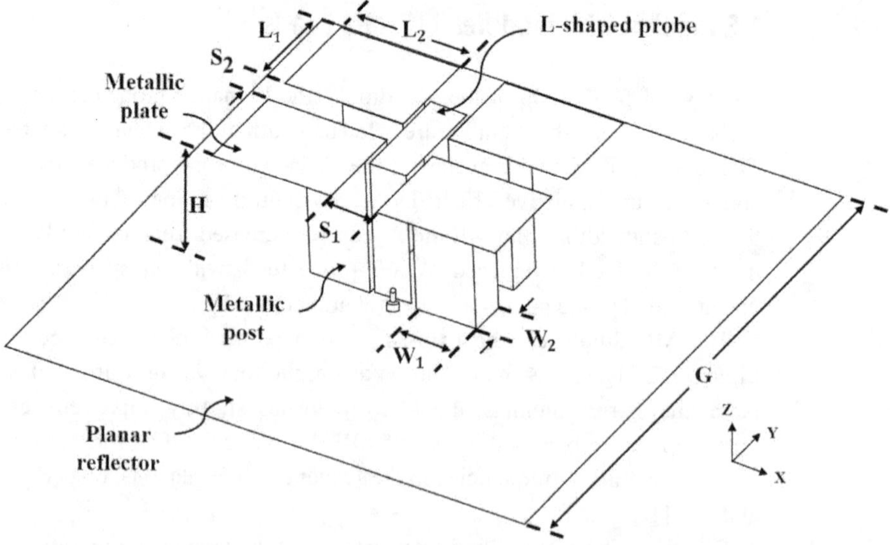

Figure 3.3 Magnetoelectric dipole with modified feed (L_1 = 34 mm, L_2 = 36 mm, W_1 = 18 mm, W_2 = 11 mm, S_1 = 14 mm, S_2 = 8 mm, H = 34 mm, and G = 160 mm) [2]

electrical performance of the basic ME dipole antenna. A typical design was demonstrated to have 93% impedance bandwidth (SWR < 2) operating from 1.62 to 3 GHz with 9 dBi gain [2]. This antenna structure can be easily modified for achieving circularly polarized and dual polarized radiation.

3.3 ME Dipoles That Are DC Grounded

For application in base stations of mobile communication systems, the ME dipole is required to be DC grounded. This can be achieved by replacing the two vertical side walls with two slant side walls and reducing the gap width of the planar electric dipole, so the magnetic dipole is changed into an equilateral triangular-shaped structure [3]. This allows the inner conductor of the coaxial feed to be connected to one arm of the electric dipole, while the outer conductor is in contact with the other arm of the electric dipole. The modification also reduces the height of the antenna to about 16% of wavelength in free space. With detailed dimensions of a prototype shown in Figure 3.4, the antenna exhibits 40% impedance bandwidth (SWR < 1.5) operated from 1.8 to 2.7 GHz. The antenna has a stable gain of 8.4 dBi with ±0.2 dB variation. The stability of the radiation pattern against frequency can be maintained.

For ease of fabrication, the triangular loop can be modified as a rectangular loop, as depicted in Figure 3.5 [4]. The gap width between the two arms of the electric dipole is kept at a small value. A horizontal strip is added for connecting the inner conductor of the coaxial feed to the opposite arm of the electric dipole. A typical example that has input SWR < 1.5 from 1.9 to 3.3 GHz (55% in bandwidth) was successfully designed and tested. This antenna has a projection area of about 0.52λ (width) by 0.57λ (length), where λ is the free-space wavelength referring to the center operating frequency. Due to the use of folded vertical walls, the height is reduced to 0.17λ. The average gain of the antenna is about 8.6 dBi with ±0.8 dB in variation. It was also

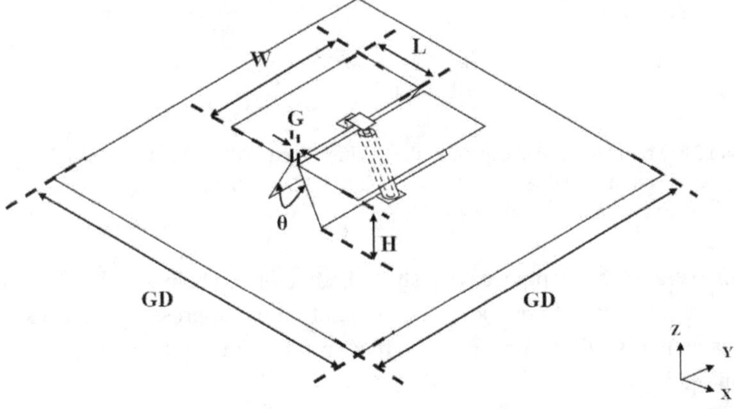

Figure 3.4 Magnetoelectric dipole with triangular loop [3]

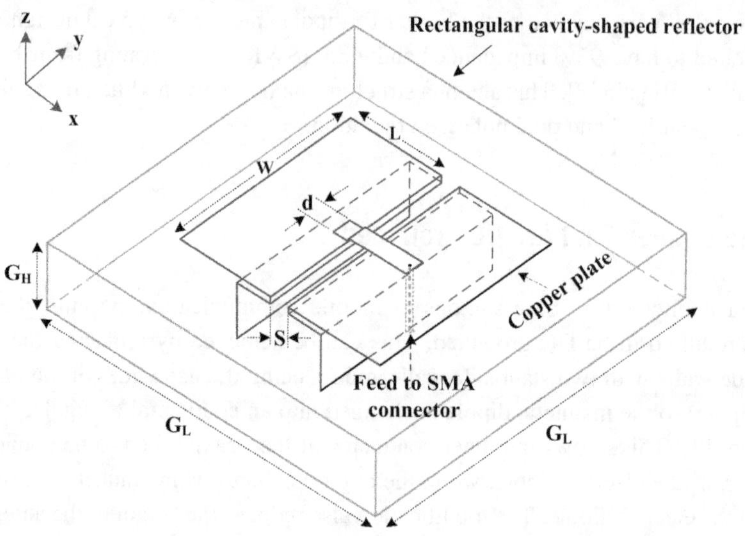

Figure 3.5 ME dipole with rectangular loop (W = 60 mm (0.52λ), L = 30 mm (0.26λ), L_1 = 1.8 mm (0.016λ), H = 19 mm (0.16λ), S = 5.4 mm (0.047λ), G_L = 112 mm (0.97λ) and G_H = 20 mm (0.17λ), where λ is the free-space wavelength referring to the center frequency.) [4]

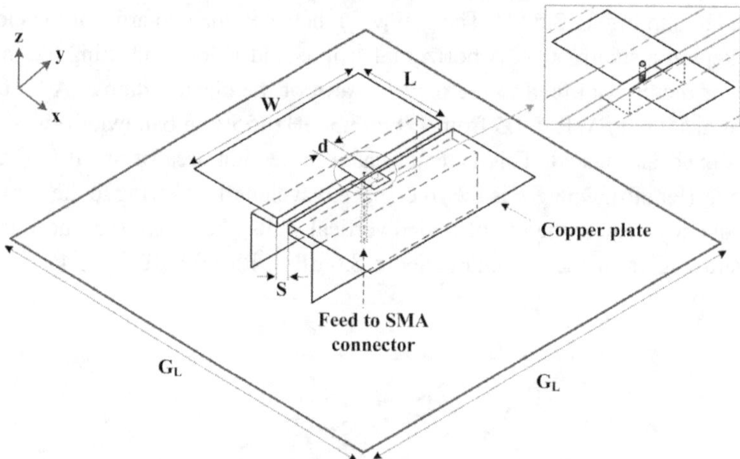

Figure 3.6 Magnetoelectric dipole with modified feed (W = 60 mm (0.48λ), L = 30 mm (0.24λ), d = 6 mm (0.048λ), and S = 4 mm (0.032λ), where λ is the free-space wavelength referring to the center frequency.) [5]

demonstrated that using a cavity-shaped reflector can help to reduce the gain variation over the operating frequency band in addition to suppressing the back radiation. As the antenna is DC grounded, it is suitable for application in base stations of mobile communications.

The feed structure of the antenna described earlier was modified as shown in Figure 3.6 [5]. Here, the coaxial feed is put at the center line so the radiation pattern in the E-plane can be more symmetrical compared with the previous design. The electric current is flown from the coaxial probe to the two arms of the horizontal electric dipole through two small strips, with the folded one in contact with the outer conductor of the coaxial cable and the straight one in contact with the inner conductor of the coaxial cable. The size of this antenna is almost identical to the antenna, as shown in Figure 3.5. Using the modified feed structure, the impedance bandwidth is reduced slightly to 46% with SWR < 1.5 from 1.86 to 2.96 GHz. The average gain is reduced slightly to 8.1 dBi with ±0.8 dB in variation. This version of ME dipole is also DC grounded.

3.4 Aperture-coupled ME Dipole

The probe-fed techniques are not easily realized at high frequencies. For achieving simple structure, the aperture-coupled feed was employed to excite the ME dipole. As shown in Figure 3.7, a rectangular slot is etched in the ground plane for coupling the energy from the microstrip line to the region between the two vertical walls which perform as a magnetic dipole. It was also demonstrated that the horizontal electric dipole can also be excited effectively. An additional advantage of using the aperture-coupled feed is that wider bandwidth can be achieved if the length of the slot is close to half wavelength. From the variation of input impedance against frequency [6], the lower operating frequency band is contributed by the electric dipole, the middle operating frequency band is contributed by the magnetic dipole, and the upper operating band is due to the coupling slot. Although an ME dipole operated between 2.5 and 6.2 GHz, equivalent to 85% impedance bandwidth for SWR < 2, was successfully designed, the backward radiation at upper operating frequencies is too high for practical applications. To resolve this drawback, a high-impedance

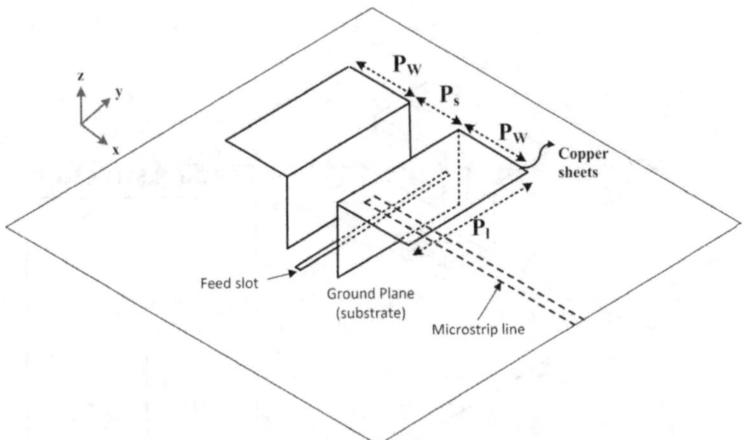

Figure 3.7 Aperture-coupled magnetoelectric dipole. Adapted from [6]

surface may be added below the microstrip line to form an electromagnetic band-gap structure for suppressing the excitation of surface waves [7].

3.5   ME Dipole with Differential Feed

Differential circuits are increasingly popular owing to their advantages of low mutual coupling between different circuit components, suppression of harmonics, low noise generation, and high linearity. For easy integration with differential RFICs (radio-frequency integrated circuits) and differential MMICs (microwave monolithic integrated circuits), differentially fed antennas are preferable as lossy on-chip or bulky off-chip baluns are not required. More importantly, differentially fed antennas have better polarization purity.

It was disclosed that the conventional single-fed ME dipole can be easily modified with a differential feed [8]. As depicted in Figure 3.8, the conventional ME dipole is excited by a rectangular loop feed which consists of two vertical strips and one horizontal strip. The two vertical strips serve as air microstrip lines and the horizontal strip is responsible for coupling the energy to the electric and magnetic dipoles. It was demonstrated that the width of the horizontal strip can be adjusted to enhance the bandwidth. A prototype operating at around 1.5 GHz was constructed to confirm the excellent performance of the antenna. The projection area is about $0.7\lambda \times 0.5\lambda$ where λ refers to the center frequency of 1.5 GHz. This differentially fed ME dipole is slightly larger than the conventional one, but the bandwidth of the antenna is increased to about 92% for SWR < 2, the cross polarization is lower than −25 dB, and the back radiation is below −19 dB. The gain is varied from 6.7 to 7.7 dBi over the operating frequencies. Symmetrical radiation patterns in both E-plane and H-plane are also found.

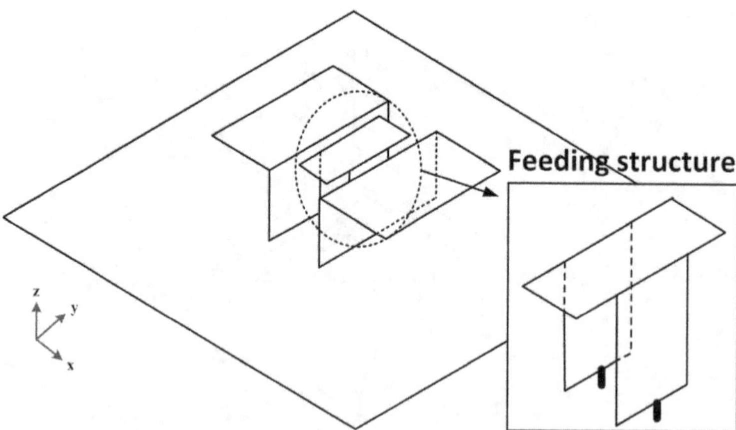

Figure 3.8 Differentially fed magnetoelectric dipole [8]

3.6   Applications of ME Dipoles in Multiple Antenna Systems

The linearly polarized ME dipole was developed for application in the 4G LTE base stations with MIMO capability. The antenna element is required to operate over a wide frequency range from 1.7 to 2.7 GHz, with interference rejection against other communication systems operating at around 2.45 GHz. This requirement was achieved by etching two face-to-face U-shaped slots on each arm of the horizontal planar electric dipole, as depicted in Figure 3.9. With appropriate dimensions [9], a wideband unidirectional antenna element operating from 1.825 to 2.925 GHz with a band-notched range from 2.39 to 2.52 GHz was demonstrated.

For testing the performance of the antenna element in an MIMO environment, a 2×3 antenna array of six band-notched ME dipoles was considered. As shown in Figure 3.10, adjacent elements are oriented perpendicular to each other. Over 30 dB isolation between adjacent antenna elements was found if the element spacing was chosen as 140 mm, about one free-space wavelength referring to the center frequency. The calculated envelope correlation coefficients between the antenna elements are

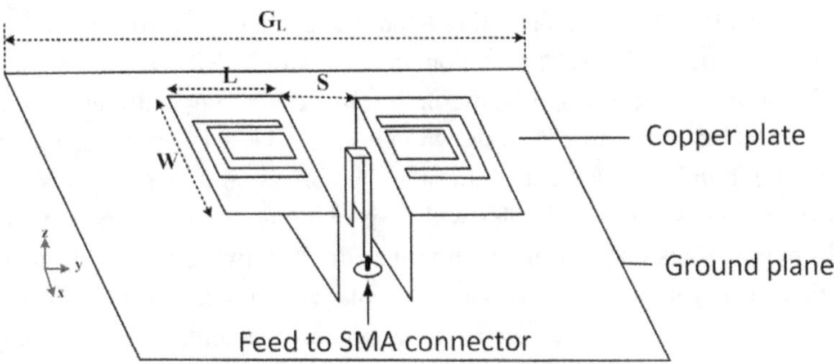

Figure 3.9 Geometry of a band-notched magnetoelectric dipole [9]

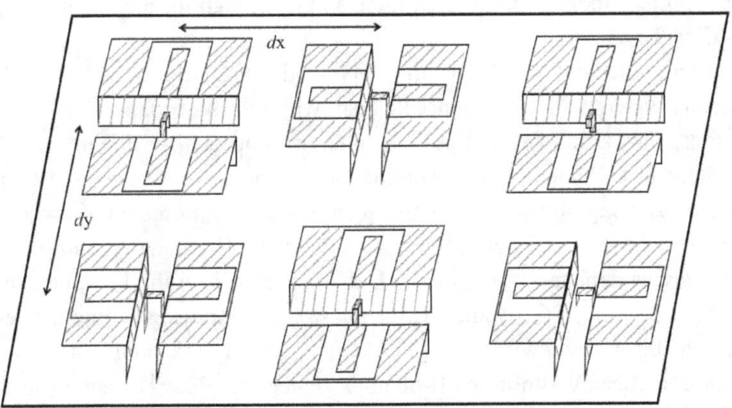

Figure 3.10 Magnetoelectric dipoles in MIMO environment [9]

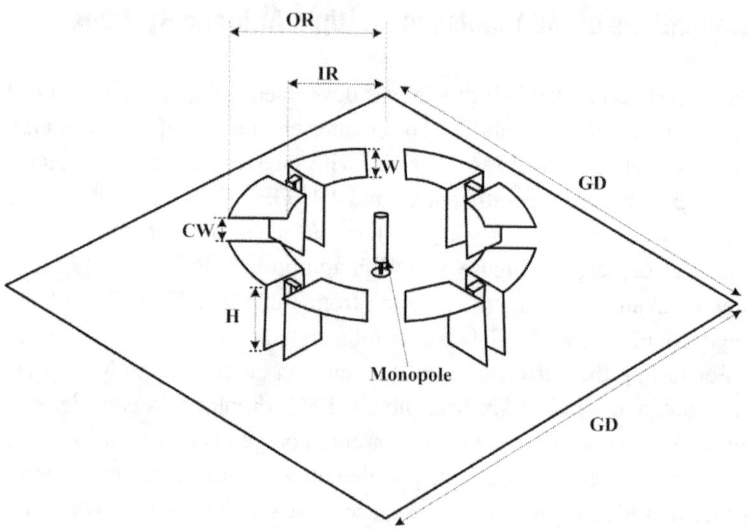

Figure 3.11 A 4-port diversity antenna with pattern and polarization diversities [10]

lower than 0.01 over the operating bands. These results verify the feasibility of this antenna element as LTE base-station antenna with MIMO capability.

Using four narrow strip ME dipoles arranged in a ring configuration together with a vertical electric monopole as shown in Figure 3.11, a 4-port wideband antenna operating at around 2.4 GHz with pattern and polarization diversities was proposed [10]. By exciting the four ME dipoles with a special microstrip feed network consisting of two hybrid rings and a Wilkinson power divider, two orthogonal broadside modes with 11 dBi gain and one horizontally polarized conical mode with 6 dBi gain can be produced. The vertical electric monopole at the center was originally narrow in bandwidth, but its bandwidth was also increased with the presence of the four ME dipoles. The four radiating modes exhibit wideband performance with stable radiation patterns and gain over 22% overlapping bandwidth. The isolation between any two input ports is more than 26 dB, which is high enough for the antenna to perform well in MIMO systems.

A three-element ME dipole array operated at around 2 GHz was used to realize an antenna system with reconfigurable beamwidth for base stations in mobile communications. This kind of reconfigurable antenna is needed for dynamic control of traffic capacity in cellular systems. Due to the use of narrow strips and half-wavelength element spacing, the entire antenna occupies a projection area of about 0.6λ by 1.2λ, where λ is the wavelength referring to the center frequency. With the alignment of the array as depicted in Figure 3.12, the beamwidth in the H-plane can be switched between 37° and 136° by altering the phase of the input signal into the central antenna element through a switchable microstrip feed network mounted below the ground plane. The beamwidth in the E-plane is centered at 72°. The achievable overlapping bandwidth is about 15%, over which the beamwidth has only 2° variation in the E-plane and H-plane.

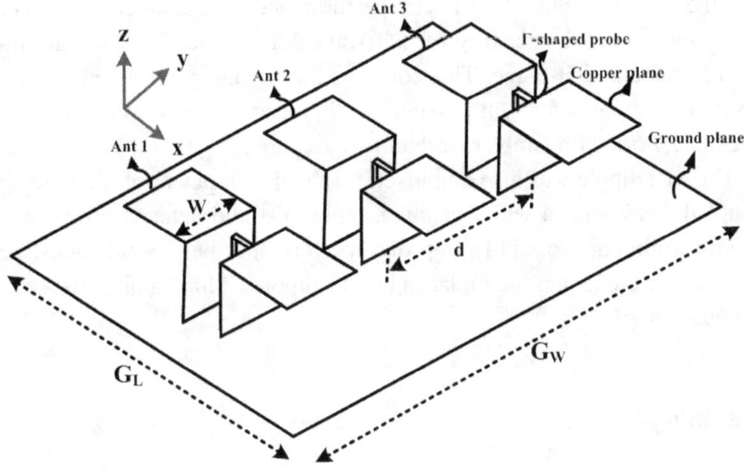

Figure 3.12 A three-element ME dipole with beamwidth reconfiguration capability [11]

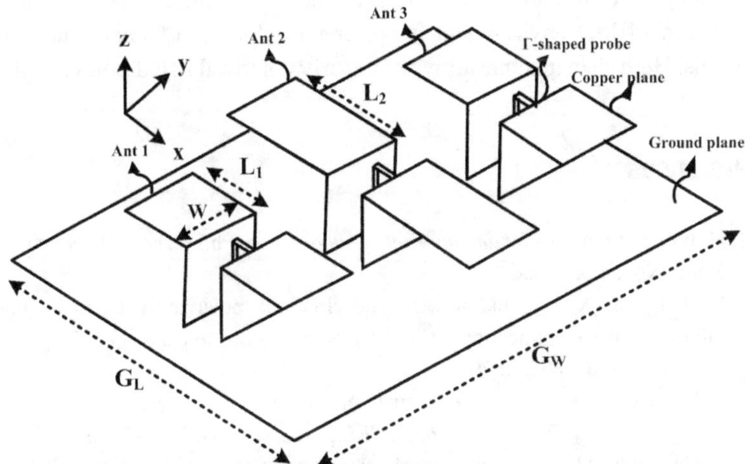

Figure 3.13 A three-element ME dipole linear array with element-level pattern diversity. Adapted from [12]

Based on the technique of element-level pattern diversity, a three-element linear array of ME dipole was synthesized and tested [12]. As shown in Figure 3.13, the element at the center has two longer horizontal arms which operate as a one-and-a half wave electric dipole. The other two elements are conventional ME dipoles with the two horizontal arms operating as a half-wave electric dipole. The magnetic dipole parts of the three ME dipoles are the same. This technique provides the possibility of E-plane radiation pattern synthesis through the superposition of the radiation patterns of different modes with appropriate weighting. The H-plane radiation pattern is simply described by the conventional array factor with the same weighting. A prototype operating at around 2.4 GHz was fabricated. The measured bandwidth is about 12.5%

and the gain is about 11 dBi. The performance of this antenna is comparable to a 3 × 3 two-dimensional array of half-wave ME dipoles having high directivities in both E-plane and H-plane. This confirms the advantage of the element-level pattern diversity technique for synthesizing ME dipole arrays, as fewer antenna elements and a feed network with simpler structure are required.

The ME dipole was also employed for the development of ultrawide band, tightly coupled array with a wide scanning angle [13] and antenna arrays with a reduced beam-forming network [14]. Improved performance has been demonstrated when the conventional antennas are replaced by ME dipoles. More applications are expected in the near future.

3.7   Summary

The development of the linearly polarized ME dipoles is reviewed. Various techniques for single-band and dual-band operation are summarized. Available feeding structures, including the differential feeds, are presented in detail. The performance of the ME dipoles in MIMO environment is presented, which is of importance in 5G mobile systems. High-gain antenna arrays of linearly polarized ME dipoles are also discussed.

References

[1] H. Wong, *A novel wideband unidirectional antenna*, Ph.D. Thesis, City University of Hong Kong, November 2006.

[2] M. J. Li and K. M. Luk, A wideband circularly polarized antenna for microwave and millimeter-wave applications, *IEEE Transactions on Antennas and Propagation*, vol. 62, 2014, no. 4, pp. 1872–1879.

[3] K. M. Luk and B. Q. Wu, The magneto-electric dipole: A new antenna element for wireless communications, *Proceedings of the IEEE*, vol. 100, 2012, no. 7, pp. 2297–2307.

[4] L. Ge and K. M. Luk, A low-profile magneto-electric dipole antenna, *IEEE Transactions on Antennas and Propagation*, vol. 60, 2012, no. 4, pp. 1684–1689.

[5] L. Ge and K. M. Luk, A magneto-electric dipole antenna with low-profile and simple structure, *IEEE Antennas and Wireless Propagation Letters*, vol. 12, 2013, pp. 140–142.

[6] X. Cui, F. Yang, M. Gao, L. Zhou, Z. Liang and F. Yan, Wideband magnetoelectric dipole antenna with microstrip line aperture-coupled excitation, *IEEE Transactions on Antennas and Propagation*, vol. 65, 2017, no. 12, pp. 7350–7354.

[7] J. Sun and K. M. Luk, Wideband magneto-electric dipole antennas for millimeter-wave applications with microstrip line feed, *Proceedings of the 2018 International Symposium on Antennas and Propagation*, Busan, South Korea, 2018.

[8] S. W. Liao, Q. Xue and J. H. Xu, A differentially fed magneto-electric dipole antenna with a simple structure, *IEEE Antennas and Propagation Magazine*, vol. 55, 2013, no. 5, pp. 74–84.

[9] J. Zhang, Y. Zang, Q Gao and C. Liang, An LTE base-station magnetoelectric dipole antenna with anti-interference characteristics and its MIMO system application, *IEEE Antennas and Wireless Propagation Letters*, vol. 14, 2015, pp. 906–909.

[10] B. Wu and K. M. Luk, A 4-port diversity antenna with high isolation for mobile communications, *IEEE Transactions on Antennas and Propagation*, vol. 59, 2011, no. 5, pp. 1660–1663.

[11] L. Ge and K. M. Luk, A three-element linear magneto-electric dipole array with beam-width reconfiguration, *IEEE Antennas and Wireless Propagation Letters*, vol. 14, 2015, pp. 28–31.

[12] D. Hua, W. Wu and D. G. Fang, The synthesis of a magneto-electric dipole linear array antenna using the element-level pattern diversity (ELPD) technique, *IEEE Antennas and Wireless Propagation Letters*, vol. 17, 2018, no. 6, pp. 1069–1072.

[13] S. M. Moghaddam, A. U. Zaman, J. Yang and A. A. Glazunov, Ultrawide band tightly-coupled aperture magneto-electric dipole array over 20–40 GHz, *Proceedings of the 13th European Conference on Antennas and Propagation*, 2019.

[14] S. Kaddour, J. Milbrandt, C. Menudier, M. Thevenot, P. Pouliguen, P. Potier and M. Romier, Performances of magneto-electric dipoles in an antennas array with a reduced beam forming network, *Proceedings of the 16th European Radar Conference*, 2019, pp. 385–388.

4 Dual-polarized and Circularly Polarized Magnetoelectric Dipoles

4.1 Introduction

Classical antennas have a single port for signal input/output and the radiated waves are linearly polarized. In many modern wireless communications, dual-polarized or circularly polarized antennas can be used to achieve better system performance. Nowadays, dual linearly polarized antennas are widely adopted in 3G and 4G mobile communications for improving the reception quality of the systems through the polarization diversity technique. Dual-polarized antennas are important for the development of massive MIMO antenna arrays for 5G mobile communication base stations operating at the sub-6 GHz band. 5G mobile phones with millimeter-wave channels also need dual-polarized printed antennas to radiate steerable directive beams for increasing the distance of communications. On the other hand, circularly polarized antennas are commonly used in satellite communications. Due to the significant advancement in satellite navigation systems, there is a big demand for low-profile wideband circularly polarized antennas, which can be used to serve all the four global navigation systems, including GPS, GLONASS, Galileo, and BEIDOU.

In this chapter, various feeding techniques and antenna structures for achieving dual-polarized and circularly polarized ME dipoles will be reviewed. Since some circularly polarized ME dipoles can be developed from dual-polarized ME dipoles, these two classes of ME dipoles are considered and reviewed together here.

4.2 Dual-polarized ME Dipoles

Conventionally, a dual-polarized antenna refers to an antenna consisting of two colocated antennas with orthogonal polarizations. The antenna therefore has two input/output ports and it can be a dual-linearly polarized antenna or a dual-circularly polarized antenna. Most of the dual-polarized ME dipole designs available in the literature are linearly polarized, and most of the authors just called their designs as dual-polarized ME dipoles.

The first dual-polarized ME dipole was disclosed in a US patent granted in 2006 [1]. As depicted in Figure 4.1, the horizontal crossed dipoles are implemented by 4 square patches, each of which is supported by two vertical side walls located under the inner edges of the patch. Each pair of parallel side walls serves as a magnetic

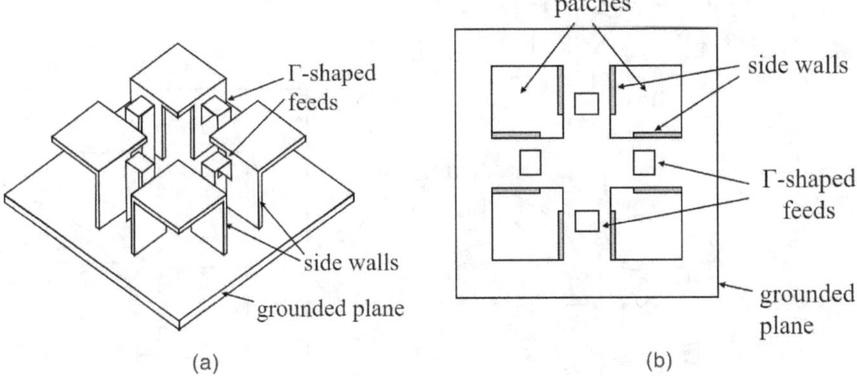

Figure 4.1 The first dual-polarized ME dipole. (a) Perspective view and (b) Top view. Adapted from [1]

dipole. Therefore, for each polarization, we can interpret the radiation coming from a two-element ME dipole array. For achieving good impedance matching, the width of the two side walls should be narrower than the width of the patch and the side walls should not be touched each other. In this design, each radiating mode is excited by two Γ-shaped feeds, so an additional power combiner or splitter is needed to be mounted below the grounded plane. The height of this design can be reduced if the regions between the parallel side walls are filled with dielectric materials [2], which will be discussed in Chapter 5.

4.2.1 Dual-polarized ME Dipole with Simple Feeds

With the objective to achieve one input/output port for each polarization, a dual-polarized ME dipole with simple feeds was reported in 2019 [3], as shown in Figure 4.2. In this design, only two Γ-shaped feeds are needed. Each Γ-shaped feed excites not only a diagonal pair of horizontal patches forming an electric dipole but also the cross-shaped apertures which radiate as magnetic currents. Each aperture can be interpreted as the open end of a vertically oriented quarter-wave microstrip patch antenna, so its characteristics can be predicted by simply analytical formulas [4]. The two side walls supporting each horizontal patch are connected to each other in this structure, which is different from the design described in the previous section.

In contrast with the basic ME dipole design with single linearly polarized radiation, the two vertical portions of the Γ-shaped feeds are located outside the central region bounded by the vertical walls, with the longer vertical portion serving as a tapered transmission line with a cornered grounding structure and the shorter vertical portion serving as an open-circuited transmission line. By drilling two small triangular holes at the corners between the horizontal patches and vertical side walls, the Γ-feeds are not touching any parts of the ME dipole structure. This new feed arrangement can help to reduce the coupling between the two feeds dramatically. Since the two feeds cannot be touching each other for good isolation, the heights of them should not be identical.

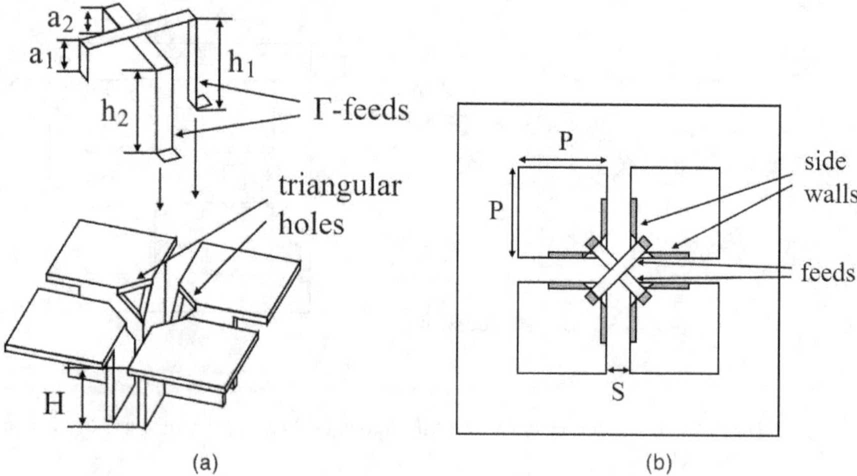

Figure 4.2 Dual-polarized ME dipole with simple feeds. (a) Perspective view and (b) Top view. (H = 28 mm ($0.23\lambda_0$), P = 29.2 mm ($0.23\lambda_0$), S = 6.2 mm ($0.05\lambda_0$), h_1 = 27 mm ($0.23\lambda_0$), h_2 = 21 mm ($0.18\lambda_0$), a_1 = 4.5 mm ($0.04\lambda_0$), a_2 = 4.0 mm ($0.03\lambda_0$)). Adapted from [3]

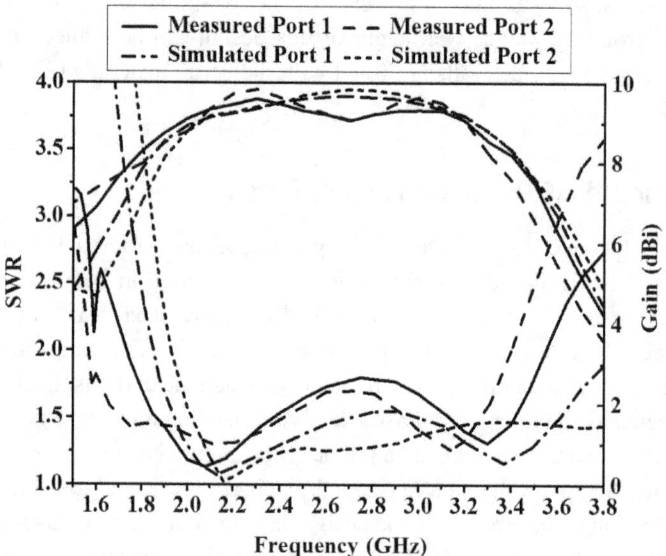

Figure 4.3 Simulated and measured data of SWR and gain (Figure 5 of [3])

After detailed parametric studies for a dual-polarized ME dipole operating at around 2.45 GHz, the height for one feed was selected as $0.23\lambda_0$ (same as the height of the antenna), and the second one was $0.18\lambda_0$, where λ_0 refers to the center frequency. Due to this difference in height, the characteristics of the two input/output ports are not the same, but the result is still promising. As shown in Figure 4.3, the antenna can be operated for both polarizations from 1.72 to 3.41 GHz with SWR < 2, effectively

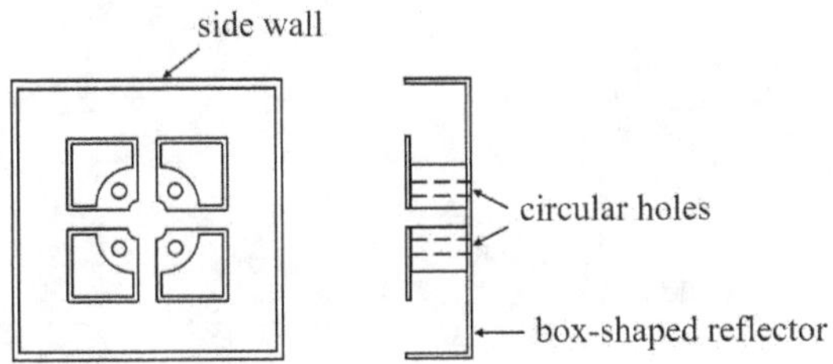

Figure 4.4 A dual-polarized base station antenna widely used in mobile communication industry. Adapted from [5]

67% impedance bandwidth, with more than 36 dB port isolation over the operating frequencies. This was achieved by offsetting the resonant frequencies of the electric dipole and magnetic dipole modes appropriately. With a projection area of about $0.51\lambda_o \times 0.51\lambda_o$, the antenna exhibits over 8 dBi in gain over the operating frequencies. Having similar advantages of the linearly polarized ME dipole, the beamwidth and gain of this dual-polarized ME dipole are not varied too much over the operating frequencies and the back lobes are lower than −20 dB over the operating frequencies even with a small ground plane. The cross-polarization levels are below −20 dB over such a wide operating frequency range, which is much better than most dual-polarized microstrip patch antennas.

A wideband dual-polarized antenna design, as depicted in Figure 4.4, was found in a US patent, which was filed in late 2005 and granted in 2007 [5], after we published the dual-polarized ME dipole structure with simple feeds in 2009. The antenna has been widely used to design many base station antennas and arrays in the mobile communication industry globally. Its structure is quite similar to the dual-polarized ME dipole, as shown in Figure 4.2. The difference is that the horizontal part consists of four loops instead of four patches. It is therefore a complementary antenna in nature. However, the gap width between neighboring loops is very small, so the magnetic dipole may not be strongly excited. Also, the use of lightweight loops instead of square patches has a negative effect on the antenna bandwidth. Some recent designs achieved improved performance by replacing the loops with patches [6] and [7].

4.2.2 Dual-polarized ME Dipoles with Modified Feeds – Version 1

For ease of installation of the antenna feeds, the whole feed structure was put inside the region bounded by the vertical walls. As shown in Figure 4.5, this was made possible by removing a small triangular portion of the inner corner of each square patch and locating a vertical wall under the new edge of each patch. In doing so, each patch becomes an equilateral pentagon. The port isolation of this design is slightly reduced in comparison to the original structure, but it has been successfully applied for designing base station antennas for mobile communications [8, 9, 10]. In these designs, a box-shaped reflector instead of a planar grounded reflector was found useful for improving the front-to-back ratio of the ME dipoles. The patch shape was modified slightly to enhance the overall performance of the antenna.

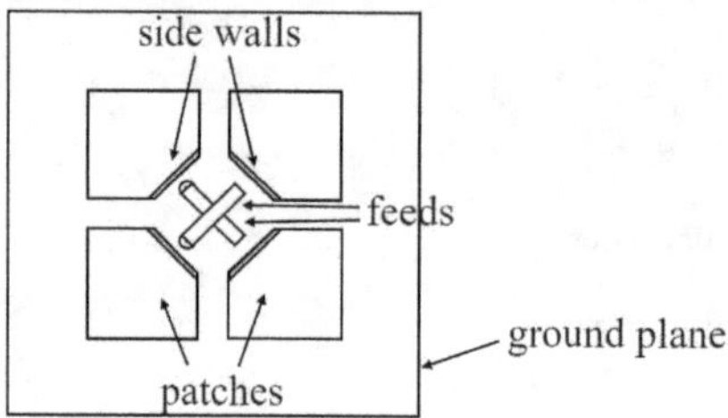

Figure 4.5 Dual-polarized ME dipole with modified feeds – version 1. Adapted from [8, 9, 10]

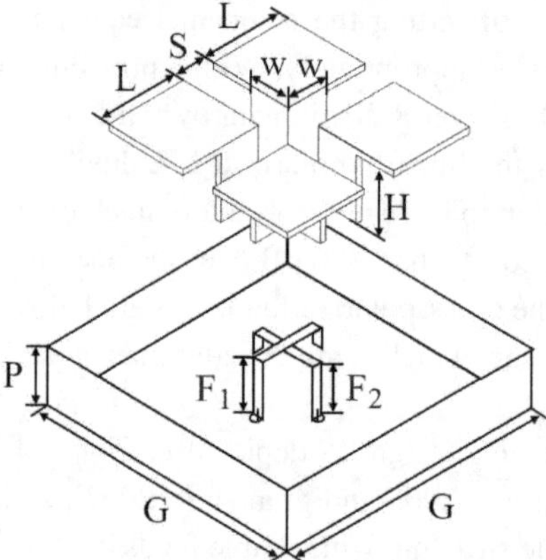

Figure 4.6 Dual-polarized ME dipole with modified feeds – version 2 (L = 27 mm (0.2λ_0), S = 6.5 mm (0.05λ_0), W = 14 mm (0.10λ_0), H = 34 mm (0.25λ_0), F$_1$ = 26 mm (0.19λ_0), F$_2$ = 23.5 mm (0.17λ_0), G = 110 mm (0.80λ_0)). Adapted from [12]

4.2.3   Dual-polarized ME Dipoles with Modified Feeds – Version 2

The linearly polarized ME dipole [11] can be easily extended for dual polarizations [12] as shown in Figure 4.6. If the designs proposed in the last section are designated as a ±45° dual-polarized radiating element, this design can also have the same designation if the dipole structure is rotated by 45°. The dipole structure of this ME dipole is identical to that in Figure 4.2, with each horizontal patch supported by a corner vertical side wall. Although the Γ-shaped feeds are located entirely inside the region confined by the four corner vertical side walls for ease of installation, the isolation can be more than 30 dB if the geometric parameters of the dipole structure and the feeds are chosen appropriately. The heights of the two feeds should be slightly different to avoid touching each other for achieving high port isolation.

A prototype with dimensions given in the caption of Figure 4.6 was built for testing. The projection area of the antenna is about 0.45λ_o × 0.45λ_o, where λ_o refers to the center operating frequency. A box-shaped reflector was selected with bottom dimensions of 0.8λ_o × 0.8λ_o. With the height of the side walls of the box reflector slightly higher

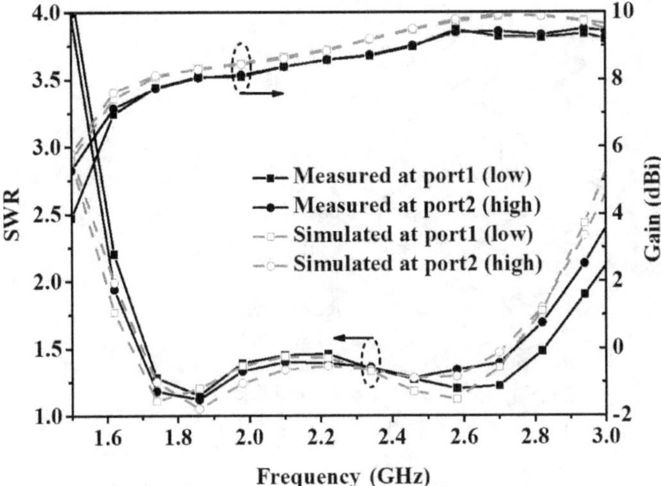

Figure 4.7 Simulated and measured data of SWR and gain of dual-polarized ME dipole (Figure 7 of [12])

than that of the ME dipole, the two input/output ports have a common impedance bandwidth of 48%, with SWR < 1.5. The gain increases almost linearly from about 7.6 to 9.3 dBi when the operating frequency is increased from about 1.69 to 2.76 GHz for both input/output ports, as shown in Figure 4.7. This behavior is different from the structure with a planar reflector that gives a more stable gain value over the operating frequencies. The beamwidth decreases gradually with increasing operating frequency in both polarizations. With the use of the box-shaped reflector, the advantages are that more than a 30 dB front-to-back ratio and 30 dB port isolation can be achieved. The antenna can be used for realizing a MIMO cube with pattern diversities [13].

4.2.4 Aperture-coupled Dual-polarized ME Dipoles

The aperture coupling technique is widely used to excite microstrip antennas. In general, wider bandwidth can be achieved with the use of this method as the aperture can also be radiated at appropriate frequencies. Conceptually, the antenna feed should be placed at the junction between the planar electric dipole and the quarter-wave patch antenna so the aperture feed is not suitable for the ME dipole.

The first ME dual-polarized aperture-coupled ME dipole with good performance was successfully developed by Yujian Li [14]. The antenna is operated at millimeter waves and its detailed design will be described in Chapter 7. The achievement motivated the development of aperture-coupled dual-polarized ME dipoles operating at lower microwave frequencies. As depicted in Figure 4.8, a dual-polarized ME dipole with a cross-shaped slot in the ground plane was proposed. In this design, the inner corners of the four horizontal patches are needed to be connected; otherwise, good impedance matching cannot be obtained. It was demonstrated that for achieving high isolation (>30 dB) between the two input/output ports, the substrate-integrated

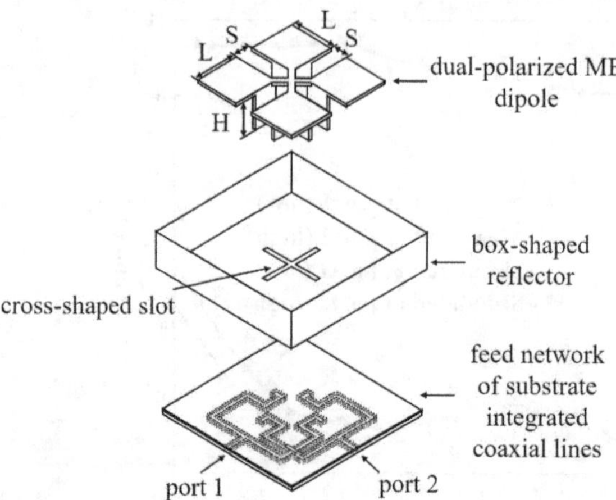

Figure 4.8 Aperture-coupled ME dipole with substrate-integrated coaxial line feed network (L = 30 mm, S = 5.8 mm, H = 36 mm). Adapted from [15]

coaxial lines can be employed to realize the feed network which is located below the ground plane. With dimensions shown in the caption of Figure 4.8, the antenna has about 23% impedance bandwidth, with SWR < 1.5 from 2.11 to 2.67 GHz. So the bandwidth of the antenna is about half of the basic design. The projection area of the antenna is about $0.53\lambda_0 \times 0.53\lambda_0$, where λ_0 is the wavelength referring to the center frequency of 2.4 GHz. The width of the gap between each pair of vertical walls, S, is about $0.05\lambda_0$, which is smaller than that of the basic design. The back lobe is lower than −24 dB if a metallic cover is added below the feed network.

4.2.5   Differentially Fed Dual-polarized ME Dipoles

The first dual-polarized ME dipoles with differential feeds was proposed by Quan Xue and his group in 2013 [16]. A cross-shaped strip was demonstrated to be an appropriate feed for exciting the dual-polarized ME dipole with almost identical radiation patterns from the two differential input/output ports. As shown in Figure 4.9, the cross-shaped strip is located inside the cross-shaped channel bounded by the four corner walls. It is supported by four vertical strips which serve as transmission lines with 50Ω characteristic impedance, connecting the four ends of the cross-shaped strip to the four input/output ports. With dimensions given in the caption of Figure 4.9, the antenna exhibits 68% impedance bandwidth, with a differential reflection coefficient less than −10 dB from 0.95 to 1.92 GHz. If the 3-dB-gain bandwidth is required, the antenna still has a wide operating band from 1.09 to 2.08 GHz. In this design, the height of the cross-shaped strip should be slightly less than the height of the electric dipoles for achieving good impedance matching. Over 36 dB isolation between the two differential ports is also achieved.

In contrast with other ME dipoles mounted over a planar ground having nearly constant gain over a wide operating band, the gain of this differentially fed ME dipole increases almost linearly from 6.6 to 9.6 dBi when the operating frequency increases

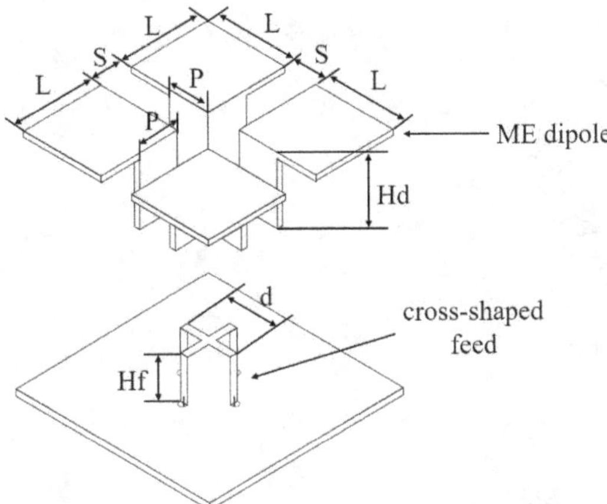

Figure 4.9 Differentially fed ME dipole (L = 53 mm (0.27λ_0), S = 16 mm (0.08λ_0), H_d = 48 mm (0.24λ_0), P = 48 mm (0.24λ_0), H_f = 32 mm (0.16λ_0), d = 34 mm (0.17λ_0)). Adapted from [16]

from 1.15 to 1.8 GHz. This antenna, in common with other ME dipoles, has low cross polarization and small variation in bandwidth, less than 10° and 4° in the H- and E-planes of each polarization, respectively. Almost identical and symmetrical radiation patterns in both polarizations and both E- and H-planes are exhibited.

4.3   Circularly Polarized ME Dipole

It is well known that using circularly polarized antenna systems can reduce polarization mismatch in satellite communications between transponders and earth stations, and cellular communications between mobile users and base stations in urban environment. Although there are many circularly polarized antenna elements that have been proposed in the literature based on electric dipoles and slots, they have their limitations in practice, such as insufficient bandwidth or high back radiation. This gives us motivation to develop high-performance circularly polarized magnetoelectric dipoles.

A circularly polarized radiation has two perpendicular field components with identical magnitude and 90° phase difference. This can be achieved by employing a single feed or double feed approach for many antennas, including the microstrip antenna and dielectric resonator antenna. These two approaches have been used to design a number of practical wideband ME dipoles, as reported in the literature.

4.3.1   Single-Feed Circularly Polarized ME Dipole – Version 1

The first circularly polarized ME dipole was developed by modifying the linearly polarized ME dipole structure as proposed by Li et al. [11], and using the single-fed crossed dipole approach. As depicted in Figure 4.10, the antenna is excited by a single L-shaped strip feed, with a slightly more complicated structure than the linearly polarized one. The antenna is right-hand circularly polarized (RHCP).

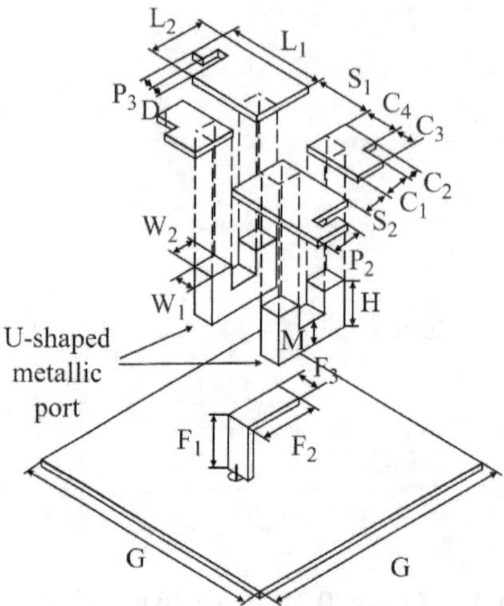

Figure 4.10 Single-feed circularly polarized ME dipole – version 1 (L_1 = 48 mm (0.33λ_0), L_2 = 28 mm (0.19λ_0), S_1 = 14 mm (0.10λ_0), S_2 = 8 mm (0.06λ_0), H = 42 mm (0.29λ_0), W_1 = 18 mm (0.13λ_0), W_2 = 11 (0.08λ_0), P_1 = 9 mm (0.06λ_0), P_2 = 16 mm (0.11λ_0), P_3 = 6 mm (0.04λ_0), C_1 = 13 mm (0.09λ_0), C_2 = 6 mm (0.04λ_0), C_3 = 18 mm (0.13λ_0), C_4 = 18 mm (0.13λ_0), M = 24 mm (0.17λ_0), F_1 = 40 mm (0.28λ_0), F_2 = 30 mm (0.21λ_0), F_3 = 12 mm (0.08λ_0), G = 160 mm (1.1λ_0)). Adapted from [11]

By removing a part of one pair of the diagonal horizontal patches, the resonant frequency of this electric dipole can be increased. Similarly, by adding an additional part to the other pair of diagonal patches, the resonant frequency of the other electric dipole can be reduced. In doing so, the phase difference between the two perpendicular field components can be 90° at frequencies between the two resonant frequencies. The removed and added parts are selected for achieving high axial-ratio bandwidth and impedance bandwidth for the crossed electric dipole, and their shapes are chosen as simple as possible for ease of fabrication.

The four horizontal patches of the ME dipole are supported by two U-shaped metallic posts instead of four metallic posts. Each U-shaped post is made of two vertical metallic posts with a smaller metallic block connecting them at the bottom. The height of the smaller metallic block can be used to increase the resonant frequency of the magnetic dipole realized by the inner regions of the two U-shaped posts. By adjusting the frequency difference between the two crossed magnetic dipoles, they can also be excited with circular polarization, which may help to enhance the axial-ratio bandwidth of the circularly polarized ME dipole.

A prototype of the proposed antenna operating at around 2 GHz was fabricated and measured. The dimensions were chosen for achieving wide bandwidth through a detailed parametric study. The total length of the L-shaped strip is 25 mm (~0.17λ_0), and the height of the antenna is 42 mm (~0.29λ_0). Other key dimensions can be found in the caption of Figure 4.10. The measured SWR of the prototype is less than 1.5 from 1.68 to 2.7 GHz, effectively 46% impedance bandwidth, and the axial ratio of the antenna is less than 3 dB from 1.69 to 2.75 GHz, effectively 47% axial-ratio bandwidth. It is remarkable that these two requirements can be achieved at the same wide frequency range. This is not an easy task even for designing other narrow-band circularly polarized antennas.

The gain of the antenna is about 8 dBic with small variation over the lower half of the operating band and then reduces gradually to 5 dBic at the highest operating frequency, probably due to the excitation of higher order modes at the upper half of the operating band. From the simulation and measurement of the radiation pattern, it is found that the patterns are not identical in the two principal planes and the left-hand circularly polarized (LHCP) radiation is quite strong in larger angle away from the broadside direction, particularly at higher frequencies. The beamwidth in one principal plane is decreased from 60° to 40° when the frequency is changed from 1.8 to 2.4 GHz. These are the drawbacks of these simple antenna structures.

4.3.2   Single-Feed Circularly Polarized ME Dipole – Version 2

The basic linearly polarized ME dipole that has a pair of rectangular patches forming the electric dipole can be modified to become a single-feed circularly polarized ME dipole. Intuitively, it was not believed to be feasible but C. H. Liang's group was able to demonstrate some exciting results [17]. As shown in Figure 4.11, a pair of rotationally symmetric patches is used as the electric dipole instead of using rectangular patches for the basic linearly polarized design. The magnetic dipole is also realized by two vertical parallel walls and a portion of the ground plane between the two parallel walls. The antenna is also fed by a Γ-shaped feed.

It was found that the modified ME dipole can radiate left-handed circularly polarized wave over 73% impedance bandwidth and 31% 3-dB axial-ratio bandwidth for a prototype with the center frequency at around 2.5 GHz. It is exciting to know that the 3-dB axial-ratio bandwidth can be enhanced to about 62% if four connected side walls

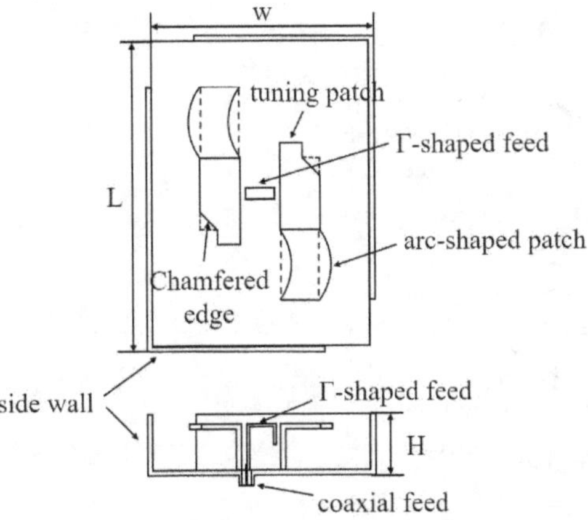

Figure 4.11 Single-feed circularly polarized ME dipole – version 2 (L = 1.6λ_0, W = 1.3λ_0, and H = 0.27λ_0). Adapted from [17]

are added at the four edges of the grounded plane, forming a box-shaped reflector. The advantage of this design is that the frequency range with axial ratio less than 3 dB is within the frequency range with SWR < 2.

It was also demonstrated that the 3-dB axial-ratio bandwidth can be further enhanced to 70% if two gaps are introduced between the side walls at two diagonal corners. Again, the advantage of this design is that the frequency range with axial ratio less than 3 dB is within the frequency range with SWR < 2. The gain varies between 7 and 9 dBi over the wide operating frequency range. This design, however, has the weakness that the axial ratio is below 3 dB over a small angular range centered at the broadside direction. Another weakness is that one dimension of the antenna is larger than one wavelength, so the antenna is not suitable for antenna array designs. Based on this, further enhancement in both impedance and axial-ratio bandwidths can be achieved by folding a portion of the patches for the electric dipole downward [18].

4.3.3   Circularly Polarized ME Dipole Excited by a Crossed Dipole

By loading a crossed dipole with an ME dipole, a circularly polarized antenna with wide bandwidth and wide beamwidth was presented by Ikmo Park's group in 2015 [19]. To this author, the antenna can also be interpreted as an ME dipole excited by a crossed dipole. The crossed dipole has a vacant-quarter rings structure to generate circularly polarized waves. The antenna is a single-feed structure with the need of a feed network and the major part of the antenna can be realized by standard PCB techniques. With four side walls surrounding the antenna, excellent performance was achieved. The antenna structure is depicted in Figure 4.12, with detailed dimensions of a prototype given in the legend of the figure. The antenna is really small in size.

The antenna exhibits 60% impedance bandwidth with SWR < 2 from 1.27 to 2.36 GHz and 27% 3-dB axial-ratio bandwidth with axial ratio less than 3 dB between 1.39 and 1.82 GHz. It was found that the circularly polarized beamwidth is larger than

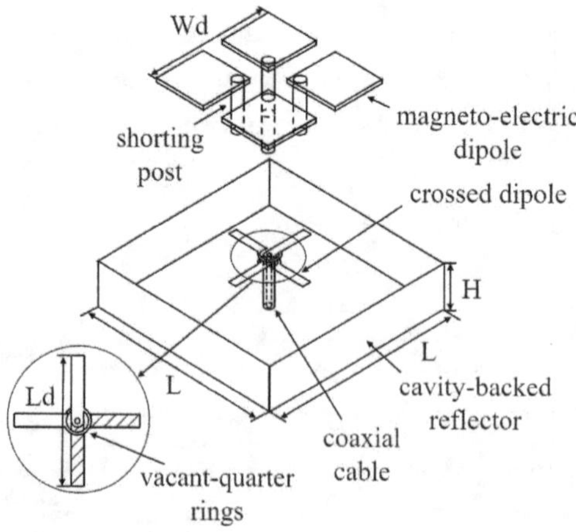

Figure 4.12 A crossed-dipole fed ME dipole. Adapted from [19]

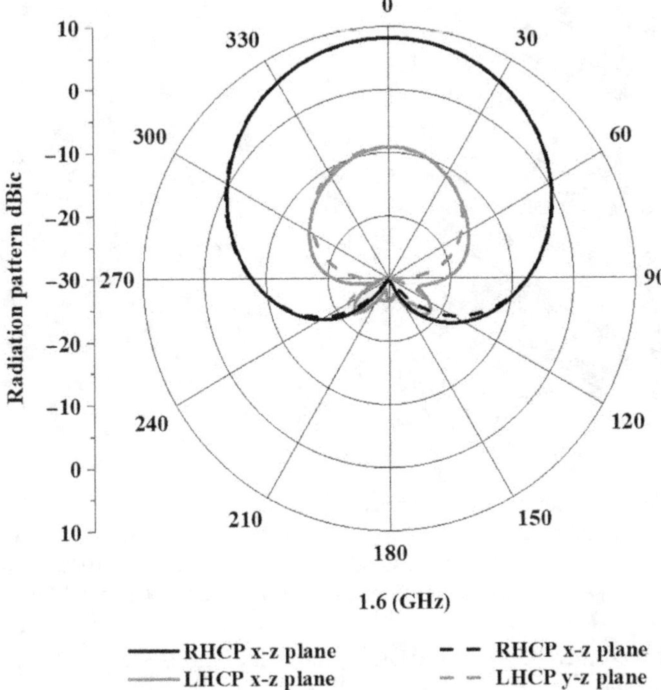

Figure 4.13 Radiation pattern of crossed-dipole fed ME dipole at center frequency of 1.6 GHz. Adapted from [19]

165° over the circularly polarized bandwidth. However, the 3-dB beamwidth is only 72°, which can be observed from Figure 4.13. Other attractive features of the antenna include stable radiation patterns and small variation in gain, low back radiation, and high radiation efficiency over the operating frequencies.

The single-feed crossed dipole itself is narrow in both impedance and axial-ratio bandwidth. When it is used as a feed for the ME dipole, the composite antenna exhibits excellent wideband characteristics which need a physical explanation, but it was not available in the published paper [19].

4.3.4 Aperture-coupled Circularly Polarized ME Dipole

As described earlier, the linearly polarized and dual-polarized ME dipoles can be excited by an aperture-coupled feed. It was demonstrated that the aperture-coupled linearly polarized ME dipole can be changed into a circularly polarized radiator if a pair of inner diagonal corners of the horizontal patches are connected together through a narrow strip, as shown in Figure 4.14 [20]. The antenna is easier to fabricate due to its simple structure. A typical design with dimensions given in the figure legend exhibits experimentally 58% impedance bandwidth with SWR < 2 from 3.48 to 6.34 GHz and 22% axial-ratio bandwidth with AR < 3 dB from 3.75 to 4.7 GHz. The gain varies from 8.7 to 9.24 dBic over the operating frequencies by measurements. Similar

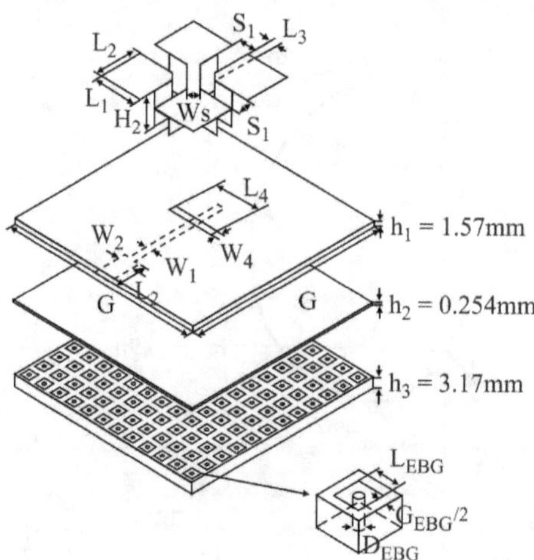

Figure 4.14 Aperture-coupled circularly polarized ME dipole ($L_2 = 16$ mm, $L_4 = 12.5$ mm, $L_5 = 2.5$ mm, $L_7 = 38$ mm, $W_1 = 3$ mm, $W_2 = 4.5$ mm, $W_3 = 26$ mm, $W_4 = 7$ mm, $W_5 = 3.54$ mm, $W_7 = 1.8$ mm, $H_2 = 18$ mm, $S_3 = 3$ mm, $G = 100$ mm, $h_1 = 1.57$ mm, $h_2 = 0.254$ mm, $h_3 = 3.17$ mm, $L_{EBG} = 8$ mm, $G_{EBG} = 1.5$ mm, $D_{EBG} = 1$ mm). Adapted from [20]

to most of the other single-feed designs, the axial ratio is small only at angles close to the boresight.

Although the antenna structure of the aperture-coupled, circularly polarized ME dipole is less complex compared with other probe-fed, circularly polarized ME dipoles, the antenna needs a mushroom layer mounted below the ground plane to achieve low back radiation and stable performance over the operating frequencies.

4.3.5 Circularly Polarized ME Dipole Based on Dual-polarized Design

In principle, all dual-polarized ME dipoles can function as circularly polarized ME dipoles if the two input/output ports are excited in equal power and 90° phase difference. One way to achieve this requirement is to incorporate a Wilkinson power divider and a 90° phase shifter. Another way is to employ a quadrature hybrid to excite the two input/output ports. The bandwidth of these additional components will determine the bandwidth of the circularly polarized ME dipole.

The wideband dual-polarized ME dipole [12], as shown in Figure 4.5, was incorporated with a wideband feed network which consists of a Wilkinson power divider and a 90° phase shifter to become a circularly polarized ME dipole. The equivalent circuit of the feed network is given in Figure 4.15. Measured results of a prototype operating at lower microwave frequencies were obtained to verify the prediction by full-wave simulation. The antenna can be operated from 1.45 to 3.05 GHz, effectively 71% bandwidth, over which SWR < 2, axial ratio < 3 dB and gain > 5 dBic. The front-to-back ratio is higher than 20 dB at frequencies above 1.5 GHz. A typical radiation pattern is shown in Figure 4.16, demonstrating the advantages of low back radiation and low cross polarization of the antenna. When the operating frequency increases

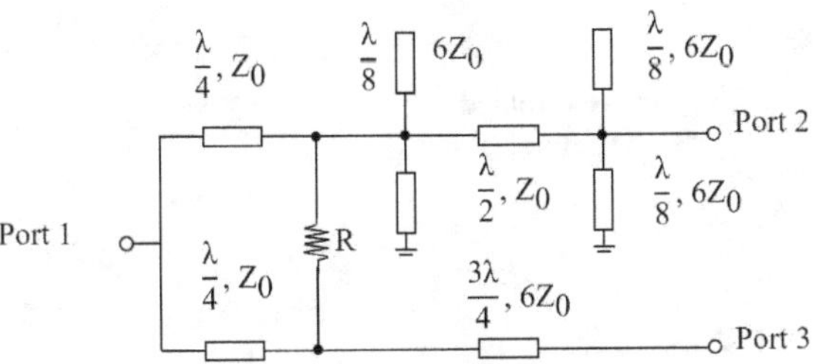

Figure 4.15 Equivalent circuit of a wideband feed network consisting of a Wilkinson power divider and a 90° phase shifter. Adapted from [12]

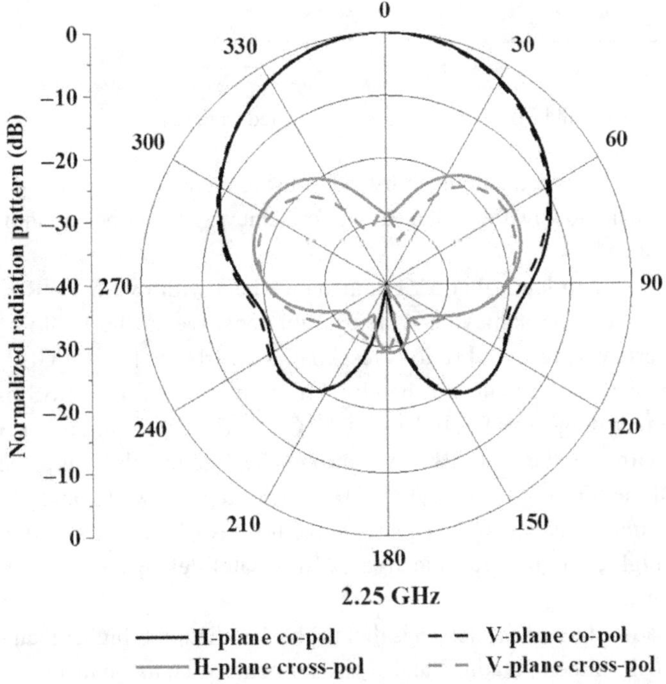

Figure 4.16 Radiation pattern of circularly polarized ME dipole at center frequency of 2.25 GHz. Adapted from [12]

from 1.6 to 3.05 GHz, the gain increases gradually from 8 to 10 dBic, and the 3-dB beamwidth decreases gradually from about 60° to 55°. This antenna was also proposed for the broadband Global Navigation Satellite Systems (GNSS) operating from 1.2 to 1.6 GHz by D C Chang's group [21].

In another design for a GNSS receiver, a wideband low-loss 3-dB directional coupler is employed to excite a dual-polarized ME dipole with two orthogonal Γ feeds locating inside the region bounded by the vertical walls [22]. As depicted in Figure 4.17, the design of the antenna is different from the conventional ones. Instead of using wide metallic plates, narrow strips of about $0.06\lambda_0$ are used to implement

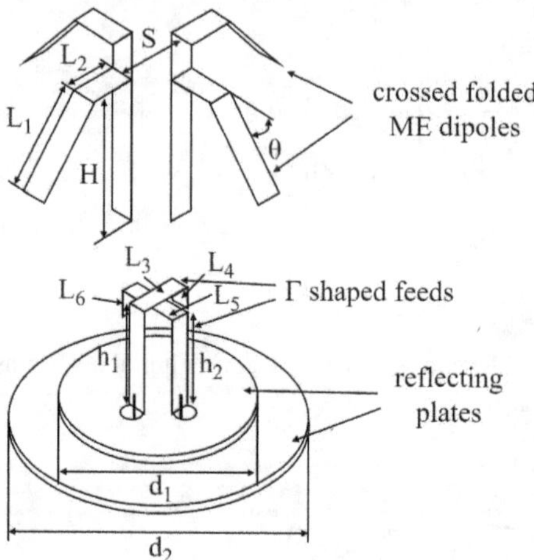

Figure 4.17 A wideband ME dipole for GNSS. Adapted from [22]

the electric and magnetic dipoles. Instead of using planar electric dipoles, the electric dipoles are bent downward with an angle for achieving wider beamwidth and smaller projection area.

It is amazing to learn that 51% impedance bandwidth with SWR < 2 from 1.1 to 1.85 GHz can be achieved, which is not sensitive to the width of the strips. More importantly, the axial ratio is less than 3 dB from 1.1 to 1.7 GHz, effectively providing over 42% axial-ratio bandwidth to cover the four GNSS bands within the frequency range from 1.163 to 1.606 GHz. The antenna has a stable gain of 5.6 dBic with less than 0.6 dB in variation. The antenna also has wide axial-ratio beamwidth, exhibiting 88° at 1.185 GHz, 91.5° at 1.227 GHz, and 98.5° at 1.575 GHz experimentally, which is an advantage for GNSS receivers as the gain at low angles is high enough to receive signals from satellites located over a wide range of directions.

One disadvantage of the design is that the back radiation is higher than −10 dB over the lower half of the operating band. This is probably due to the fact that the resonant frequencies of the electric dipole and magnetic dipole are selected far away to achieve wider bandwidth. But in doing so, the complementary effect between the electric and magnetic dipoles for reducing the back radiation cannot be achieved.

4.4   Summary

Several representatives of dual-polarized and circularly polarized ME dipoles developed by the author's group and other leading antenna experts are reviewed. The design principle of each antenna is explained. Hopefully, this chapter can provide the necessary physical insight for practicing engineers and postgraduate students who are interested to apply these basic designs for specific applications at lower microwave frequencies.

References

[1] K. M. Luk and H. Wong, *Complementary Wideband antenna*, US Patent, No.: US7,843,389 B2, Filed: March 10, 2006, Published: November 30, 2010.

[2] L. Siu, H. Wong and K. M. Luk, A dual-polarized magneto-electric dipole with dielectric loading, *IEEE Transactions on Antennas and Propagation*, vol. 57, 2009, no. 3, pp. 616–623.

[3] B. Q. Wu and K. M. Luk, A broadband dual-polarized magneto-electric dipole antenna with simple feeds, *IEEE Antennas and Wireless Propagation Letters*, vol. 8, 2009, pp. 60–63.

[4] K. F. Lee, K. M. Luk and H. W. Lai, Microstrip patch antennas, World Scientific Publishing Press, 2018.

[5] M. Boss, M. Gottl, N. Kreuzer, J. Langenberg and J. Rumold, Dual polarized antenna, US Patent, No.: US2007/0146225A1, Filed: December 28, 2005, Published: June 28, 2007.

[6] Y. Fang, Y. Sun and Z. Li, A dual-polarized magneto-electric dipole antenna with a novel feeding structure, *Proceedings of 2018 Cross Strait Quad-Regional Radio Science and Wireless Technology Conference*, Xuzhou, China, July 2018.

[7] H. Tang, X. Zong and Z. Nie, Broadband dual-polarized base station antenna for fifth-generation (5G) applications, *Sensors*, vol. 18, 2018, p. 2701.

[8] J. N. Lee, K. C. Lee, G. D. Jo, H. K. Kwon, B. S. Kang, J. H. Oh, M. D. Kim and N. H. Park, Design of the dual-polarized dipole antenna for small base station, *Proceedings of ISAP 2012*, Nagoya, pp. 1059–1062.

[9] W. Yu, Z. Zhang, A. Zhang and X. Wu, A broadband dual-polarized magneto-electric dipole antenna for 2G/3G/LTE applications, 2017 *IEEE APS Symposium Digest*, 2017.

[10] Z. Lu, Y. Sun, H. Zhu and F. Huang, A broadband ±45° dual-polarized magneto-electric dipole antenna for 2G/3G/LTE/5G/WiMAX applications, *Progress in Electromagnetics Research C*, vol. 86, 2018, pp. 153–165.

[11] M. Li and K. M. Luk, A wideband circularly polarized antenna for microwave and millimeter-wave applications, *IEEE Transactions on Antennas and Propagation*, vol. 62, 2014, no. 4, pp. 1872–1879.

[12] M. Li and K. M. Luk, Wideband magnetoelectric dipole antennas with dual polarization and circular polarization, *IEEE Antennas and Propagation Magazine*, vol. 57, 2015, no. 1, pp. 110–119.

[13] S. Chen and K. M. Luk, A dual-mode wideband MIMO cube antenna with magneto-electric dipoles, *IEEE Transactions on Antennas and Propagation*, vol. 62, 2014, no. 12, pp. 5951–5959.

[14] Y. Li and K. M. Luk, 60-GHz dual-polarized two-dimensional switch-beam wideband antenna array of aperture-coupled magneto-electric dipoles, *IEEE Transactions on Antennas and Propagation*, vol. 64, 2016, no. 2, pp. 554–563.

[15] X. Yang, L. Ge, D. Zhang and C. Y. D. Sim, Magnetoelectric dipole antenna with dual polarization and high isolation, *Hindawi Wireless Communications and Mobile Computing*, vol. 2018, article ID 4765425, 2018.

[16] Q. Xue, S. W. Liao and J. H. Xu, A differentially-driven dual-polarized magneto-electric dipole antenna, *IEEE Transactions on Antennas and Propagation*, vol. 61, 2013, no. 1, pp. 425–430.

[17] K. Kang, Y. Shi and C. H. Liang, A wideband circularly polarized magnetoelectric dipole antenna, *IEEE Antennas and Wireless Propagation Letters*, vol. 16, 2017, pp. 1647–1650.

[18] C. Q. Feng, F. S. Zhang, H. J. Zhang and J. X. Su, A single-feed circularly polarized magnetoelectric dipole antenna for wideband wireless applications, *Progress in Electromagnetics Research M*, vol. 65, 2018, pp. 1–8.

[19] S. X. Ta and I. Park, Crossed dipole loaded with magneto-electric dipole for wideband and wide-beam circularly polarized radiation, *IEEE Antennas and Wireless Propagation Letters*, vol. 14, 2015, pp. 358–361.

[20] J. Sun and K. M. Luk, Wideband linearly-polarized and circularly-polarized aperture-coupled magneto-electric dipole antennas fed by microstrip line with electromagnetic bandgap surface, *IEEE Access*, vol. 7, 2019, pp. 43084–43091.

[21] D. C. Chang and H. J. Lee, A wideband circularly polarized antenna for GNSS, *Proceedings of the IEEE 5th Asia-Pacific Conference on Antennas and Propagation (APCAP)*, 2016, pp. 343–344.

[22] E. C. Wang and L. Y. Shi, An improved wideband dipole antenna for global navigation satellite system, *IEEE Antennas and Wireless Propagation Letters*, vol. 13, 2014, pp. 1305–1308.

5   Size Reduction Techniques for Magnetoelectric Dipoles

5.1   Introduction

The basic air magnetoelectric dipole has dimensions of about $0.6\lambda_0$ in length, $0.5\lambda_0$ in width, and $0.25\lambda_0$ in height, where λ_0 is the free-space wavelength referring to the center operating frequency. It may be too thick for some compact wireless hot spots or devices. If the antenna element is used as a building block for constructing high-gain phased antenna arrays, its projection area or footprint is preferred to be less than half a wavelength referring to the highest operating frequency in order to avoid the grating lobe problem.

In this chapter, techniques for size reduction of the magnetoelectric dipole available in the literature are reviewed. The relative advantages of employing the folded patch technique, dielectric-loaded method, and the metamaterial-loaded approach are compared. Designs with single input port and differential input ports are also reviewed. Hopefully, possible new techniques will be achieved by readers after reviewing all these interesting designs.

5.2   Low-profile Linearly Polarized ME Dipoles

The first attempt [1] to reduce the height or thickness of the ME dipole is to fold the two parallel vertical walls a number of times as depicted in Figure 5.1. The idea is based on our experience in designing folded patch antennas over the years [2]. In this design, the microstrip feedline is needed to fold with the vertical wall close to it. Initially, we thought that this approach might affect the characteristics of the magnetic dipole part of the antenna and the microstrip feedline. After detailed parametric study, we achieved quite good results for this first folded ME dipole. With dimensions given in the caption of Figure 5.1, the antenna can still maintain its good performance in most of the characteristics of the antenna when the height is reduced to $0.15\lambda_0$. The impedance bandwidth is about 52% with SWR < 2 from 2.182 to 3.704 GHz, the average gain is about 8 dBi with 0.5 dB variation, the back radiation is less than -23 dB, and the cross polarization is less than -27 dB. The only performance degradation perhaps is the increase in beamwidth variation, with the E-plane beamwidth fluctuating between $61°$ and $83°$ and the H-plane beamwidth fluctuating between $55°$ and $75°$.

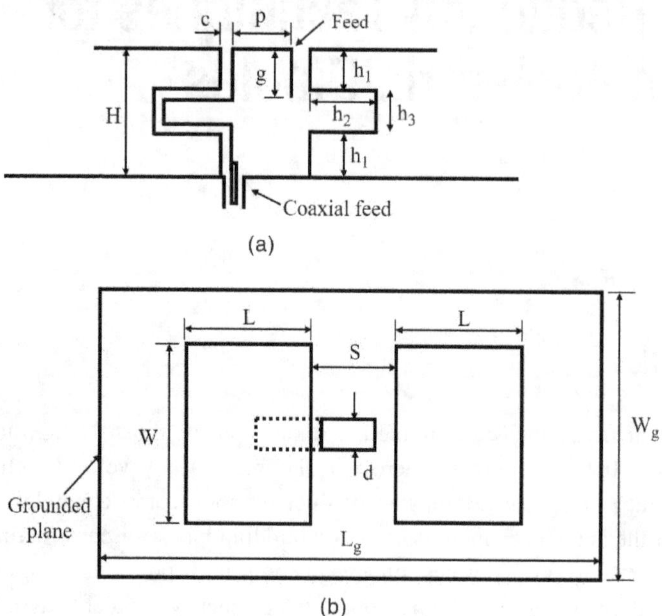

Figure 5.1 Geometry of the first folded ME dipole. (a) Side view and (b) Top view (H = 18 mm $(0.15\lambda_0)$, $h_1 = h_2 = h_3 = 6$ mm $(0.05\lambda_0)$, p = 11.5 mm $(0.1\lambda_0)$, q = 13 mm $(0.11\lambda_0)$, d = 4.9 mm $(0.04\lambda_0)$, L = 30 mm $(0.25\lambda_0)$, S = 17 mm $(0.14\lambda_0)$, W = 60 mm $(0.5\lambda_0)$, L_g = 160 mm $(1.3\lambda_0)$ and W_g = 160 mm $(1.3\lambda_0)$). Adapted from [1]

The three ME dipoles with the DC grounding feature [2–4], as depicted in Figures 3.4–3.6, are thinner than the basic ME dipole, but the thickness cannot be smaller than $0.16\lambda_0$ if we want to keep more than 40% impedance bandwidth (SWR < 1.5). These three designs are achieved by replacing the two vertical walls (magnetic dipole) by two slanted walls forming a triangular loop [2] or two folded walls forming a rectangular loop [3, 4]. The performances of these antennas are summarized in Table 5.1 for ease of reference.

One effective approach for further reducing the thickness of the ME dipole to about 10% of free-space wavelength referring to the center operating frequency is depicted in Figure 5.2. It can be seen that both the horizontal metallic sheets and vertical walls are folded two times. This approach can help to reduce the sizes of the electric dipole and the quarter-wave patch antenna. After a detailed parametric study by simulation, a compact folded ME dipole having a length of $0.5\lambda_0$ and height of $0.11\lambda_0$ was designed. With the geometric parameters given in Figure 5.2, the antenna can maintain a wide bandwidth of 44% with SWR < 2 between 1.66 and 2.60 GHz, as observed in Figure 5.3. The gain of the antenna is reduced over the operating frequencies with a peak value of 6.7 dBi. This is due to the broadening of the beamwidths in both E- and H-planes, as depicted in Figure 5.4. This compact version has the disadvantage of larger gain variation over the operating frequencies, but it has a better performance in front-to-back ratio, which is higher than 25 dB

Table 5.1 Performance of three ME dipoles with DC grounding feature

Reference	Geometry of magnetic dipole	% Bandwidth (SWR < 2)	Frequency range (GHz)	Height (H in λ_0)	Gain (dBi)
[2]	Triangular loop	40	1.83–2.71	0.161	8.4 ± 0.2
[3]	Rectangular loop	55	1.88–3.30	0.173	8.6 ± 0.8
[4]	Rectangular loop	46	1.86–2.96	0.169	8.1 ± 0.8

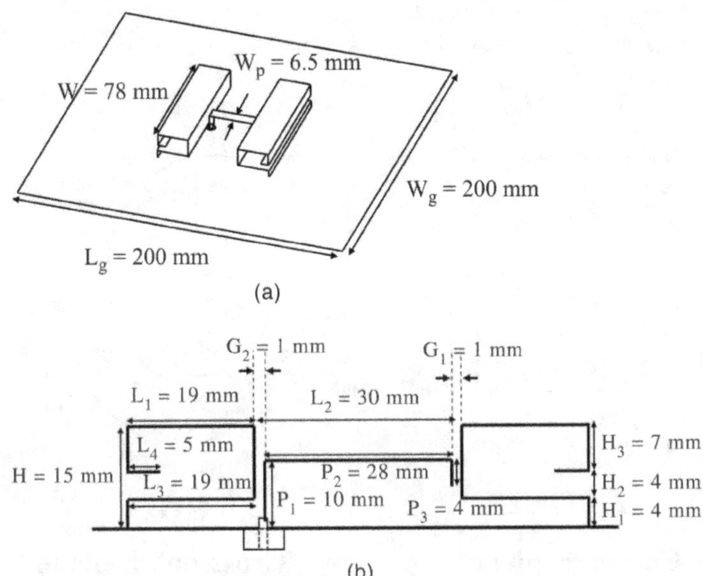

Figure 5.2 Geometry of compact folded ME dipole. (a) Perspective view and (b) Side view. Adapted from [2]

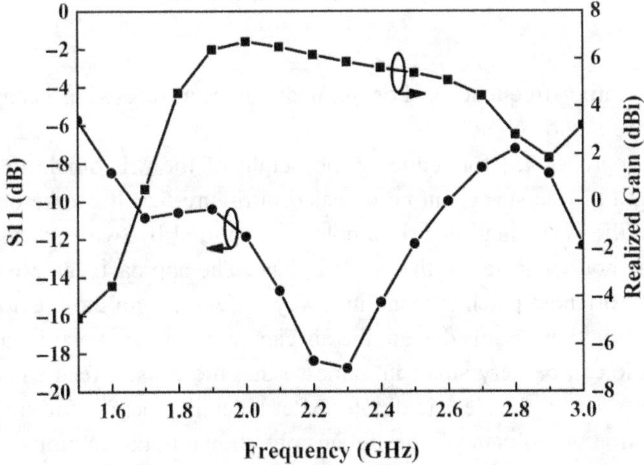

Figure 5.3 Gain and return loss of compact folded ME dipole. Adapted from [2]

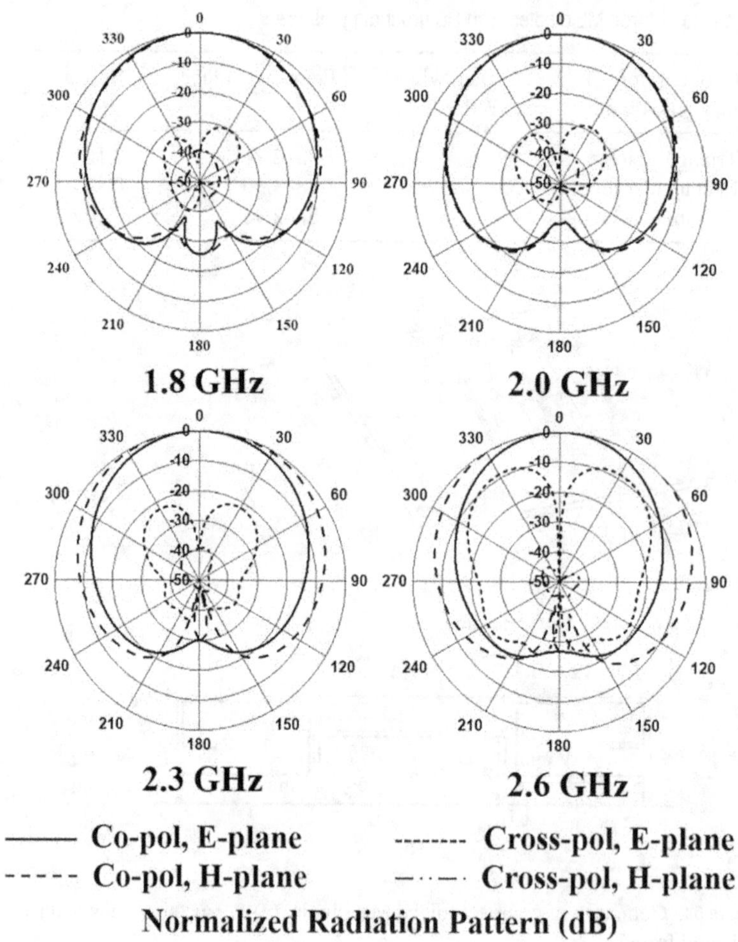

Figure 5.4 Radiation patterns of compact folded ME dipole. Adapted from [2]

over the operating frequencies. For applications requiring wide beamwidth, this design may be a good choice.

Another approach [5] for reducing the height of the ME dipole to about 10% of wavelength in free space can be revealed in Figure 5.5. It can be observed that each arm of the horizontal electric dipole is supported by two vertical walls with the outer one not connected to the ground plane. The approach can effectively fold the vertically oriented patch antenna in a way different from the previous versions for easy construction. In this design, the air gap between the two arms of the planar electric dipole can be very small in value so that the coaxial feed can be made in physical contact with the electric dipole. After a detailed parametric study, a nearly optimal solution was obtained with dimensions shown in the caption of Figure 5.5. This compact version of ME dipole has a length of $0.56\lambda_0$, a width of $0.56\lambda_0$, and a height of $0.12\lambda_0$. For achieving wideband matching performance, the width of the

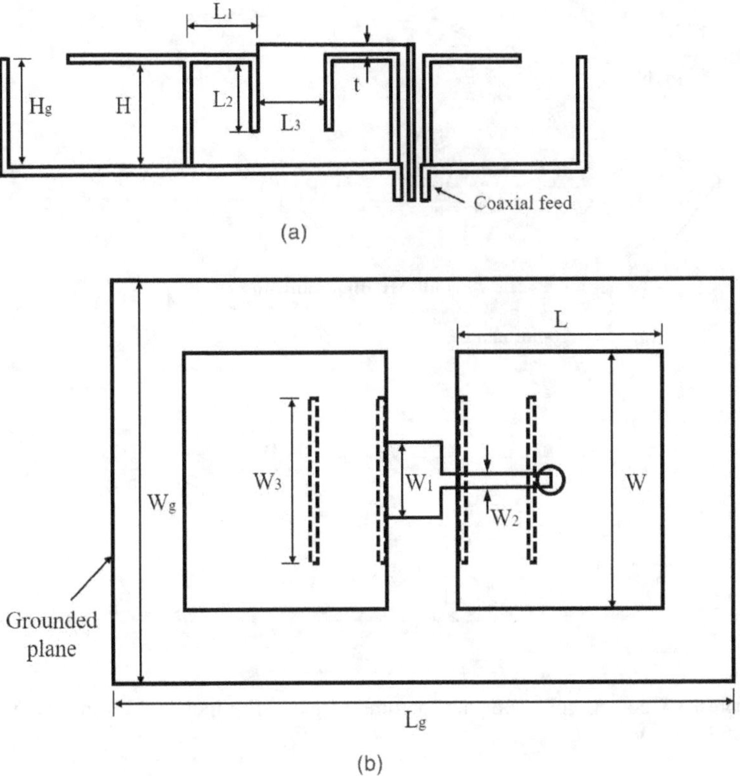

Figure 5.5 Geometry of a compact low-profile ME dipole. (a) Side view and (b) Top view ($L = 30$ mm ($0.24\lambda_0$), $L_1 = 8$ mm ($0.06\lambda_0$), $L_2 = 12$ mm ($0.1\lambda_0$), $L_3 = 10$ mm ($0.08\lambda_0$), $W = 70$ mm ($0.56\lambda_0$), $W_1 = 17$ mm ($0.14\lambda_0$), $W_2 = 2$ mm ($0.02\lambda_0$), $W_3 = 55$ mm ($0.44\lambda_0$), $L_g = 110$ mm ($0.88\lambda_0$), $H_g = 15$ mm ($0.12\lambda_0$)). Adapted from [5]

vertical walls needs to be slightly less than the width of the horizontal patches and the width W_1 of the feed line needs to be much larger than the width W_2 of the 50 Ω line. Wide impedance bandwidth of 43.6% was achieved, with SWR < 2 from 1.81 to 2.82 GHz. With the use of a cavity reflector or a box-shaped grounding structure, low back radiation of less than -15 dB over the operating frequencies was obtained, and the gain increases linearly from 8 to 10 dBi when the operating frequency increases from about 1.9 to 2.8 GHz.

The compact ME dipole described earlier can also be conveniently fed by an aperture-coupled feed, as depicted in Figure 5.6. With the selected parameters as given in the figure caption, the antenna exhibits 33% impedance bandwidth with SWR < 2 from 1.86 to 2.6 GHz. In this design, the height is reduced to $0.097\lambda_0$. As the aperture can radiate strongly to the back side of the grounded plane, a high impedance surface was added below the ground plane to achieve low back radiation, as shown in Figure 5.7. The gain is very stable, around 8 dBi by simulation, as no vertical side walls are added. The radiation pattern is shown in Figure 5.8.

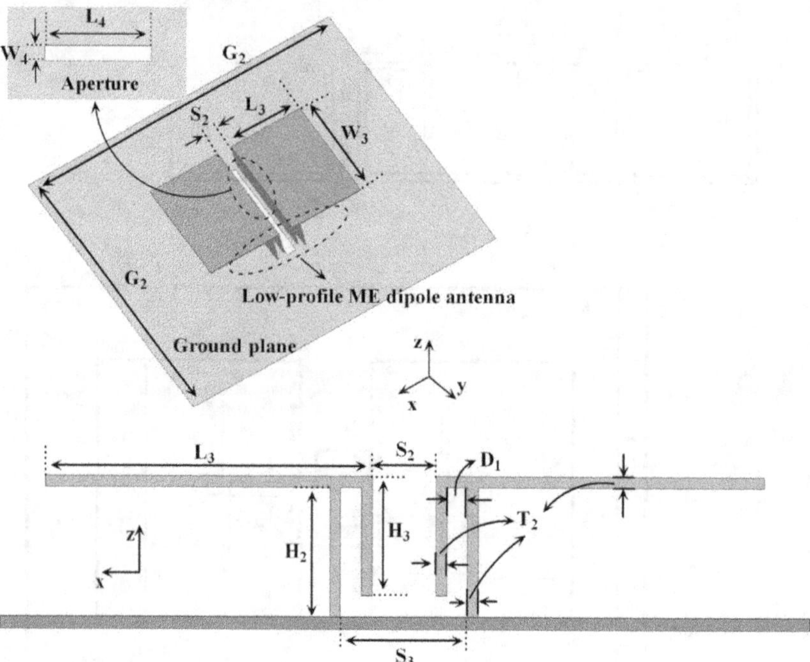

Figure 5.6 Geometry of a compact aperture-coupled ME dipole. (a) Perspective view and (b) Side view

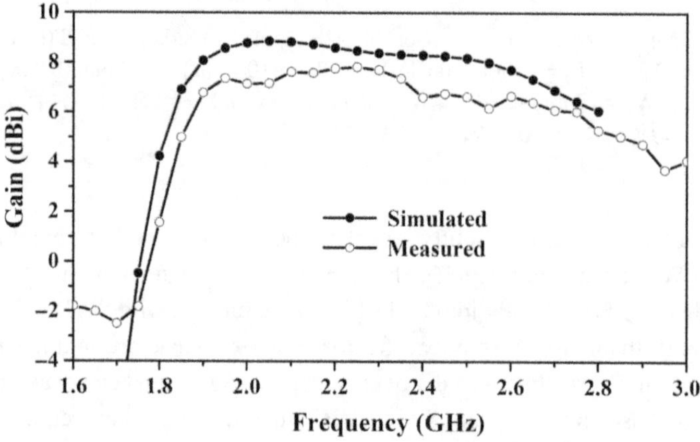

Figure 5.7 Gain of the compact aperture-coupled ME dipole

Several other methods were also proposed to reduce the height of the ME dipole in the literature, but these methods result in decreasing the bandwidth substantially. In [6], the two vertical walls for the magnetic dipole are replaced by two slant walls and the electric dipole is in the form of a bowtie shape. The height can be reduced to $0.08\lambda_0$, but the bandwidth is decreased to 20% even using a width of $0.8\lambda_0$. The advantage

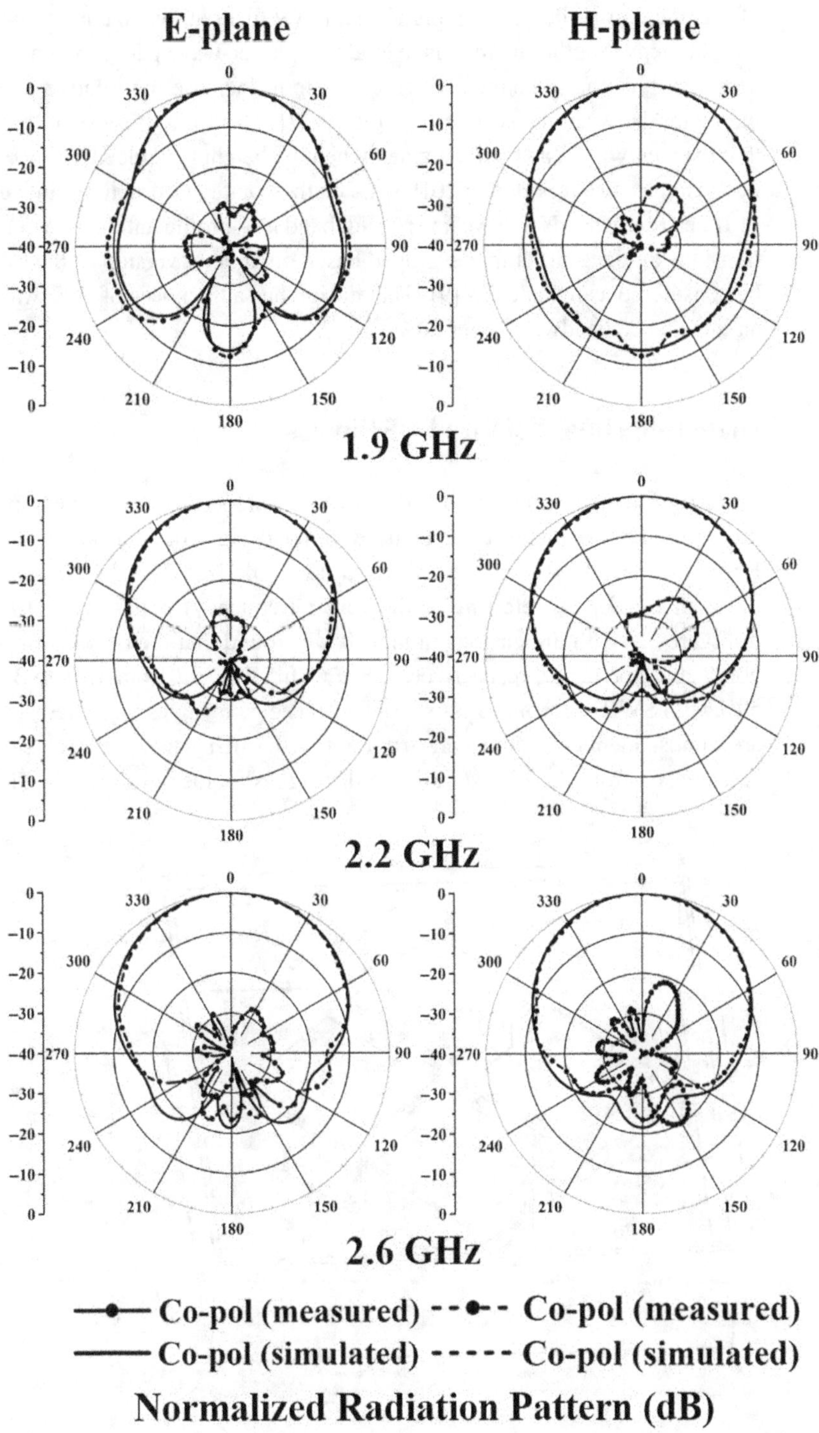

Figure 5.8 Radiation patterns of the compact aperture-coupled ME dipole

of this method is that it has a higher gain of 9 dBi but the size is too large for array environment. Another method is to load the quarter-wave patch antenna with complementary split-ring structures which can help to increase the relative permeability of the region between the two vertical planes [7]. The attractiveness of this approach is that the bandwidth can be maintained when the height is reduced up to $0.15\lambda_0$. A similar design is also reported in 2019 [8]. But these antenna structures are very complex.

If one really needs to have a very wideband low-profile antenna for certain applications, the antenna structure developed based on another version of the ME dipole can be considered as revealed in [9]. That design has a thickness of 10% wavelength and an impedance bandwidth over 50%.

5.3   Low-profile Dual-polarized ME Dipoles

The height reduction techniques that were developed for the linearly polarized ME dipoles as described before can be used for the designs of dual-polarized ME dipoles. Based on the folded patch technique, a dual-polarized ME dipole with a thickness of about $0.15\lambda_0$ (λ_0 referring to the center frequency) is designed [10] and shown in Figure 5.9 with dimensions in mm. It was found that the bandwidths evaluated at port 1 and port 2 are, respectively, 58% with SWR < 2 from 1.71 to 3.13 GHz and 59% with SWR < 2 from 1.73 to 3.18 GHz. The gain values measured at the two ports are almost identical, increasing from 8 to 10.5 dBi when the operating frequency is increased from the lowest value to the highest value, as shown in Figure 5.10. If

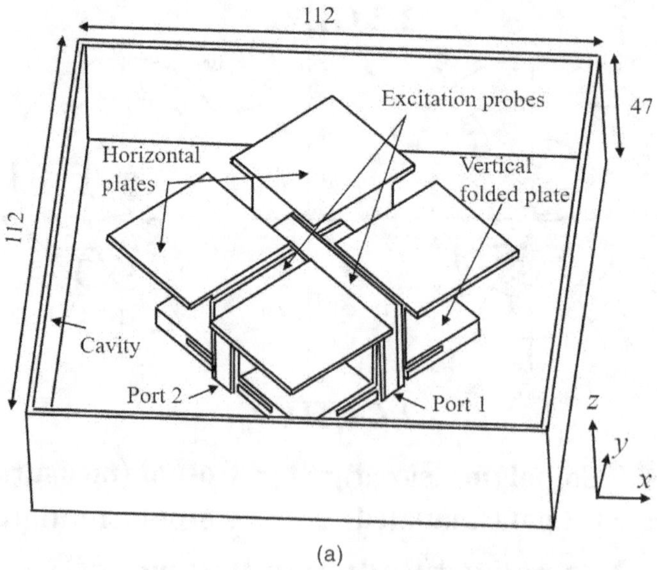

(a)

Figure 5.9 Low-profile dual-polarized folded ME dipole. (a) Perspective view of whole structure, (b) Perspective view of one quarter of the structure, (c) Feeds for port 1 and port 2 (dimensions in mm). Adapted from [10]

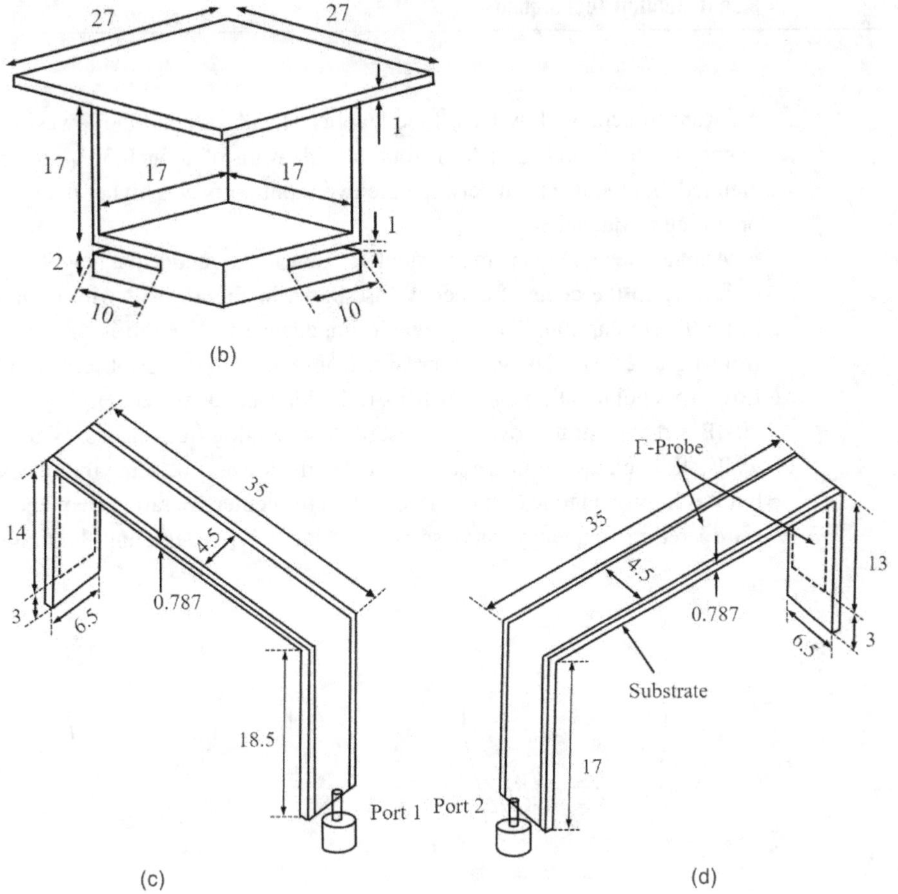

Figure 5.9 (cont.)

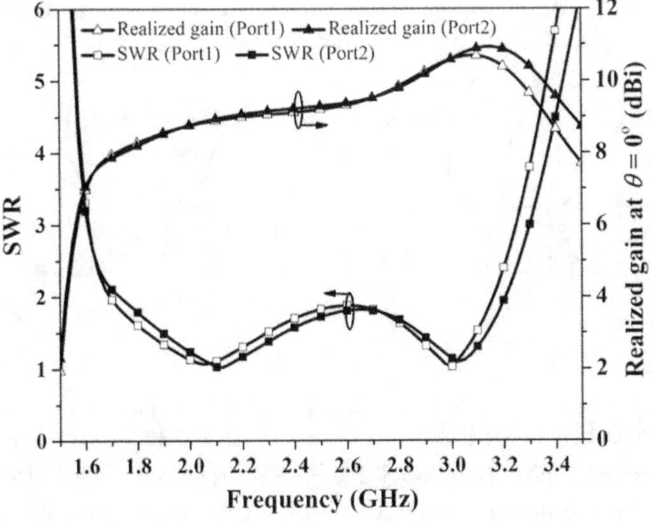

Figure 5.10 Simulated gain and SWR versus frequency of low-profile dual-polarized folded ME dipole. Adapted from [10]

we want to achieve low back lobe level especially around the lowest operating frequency as shown in Figure 5.11, four tall side walls of about $0.35\lambda_0$ are needed. It was claimed that the isolation between the two input ports was higher than 25 dB over the operating frequencies.

Another design [11] is to use slanted sidewalls to reduce the height to about $0.15\lambda_0$ (referring to the center frequency) as shown in Figure 5.12 with dimensions given in the figure caption. The achievable impedance bandwidth is 48% with SWR < 2 from 1.2 to 2 GHz. The gain increases from 8 to 10.6 dBi as observed in Figure 5.13. Low cross polarization and low back radiation were achieved. High isolation of about 30 dB is demonstrated over the most of the operating frequencies. Without using side walls, the back radiation is about -12 dB at the lowest operating frequency of 1.2 GHz but it is improved to less than -18 dB from the center operating frequency to the highest operating frequency, as observed in Figure 5.14, depicting the radiation excited

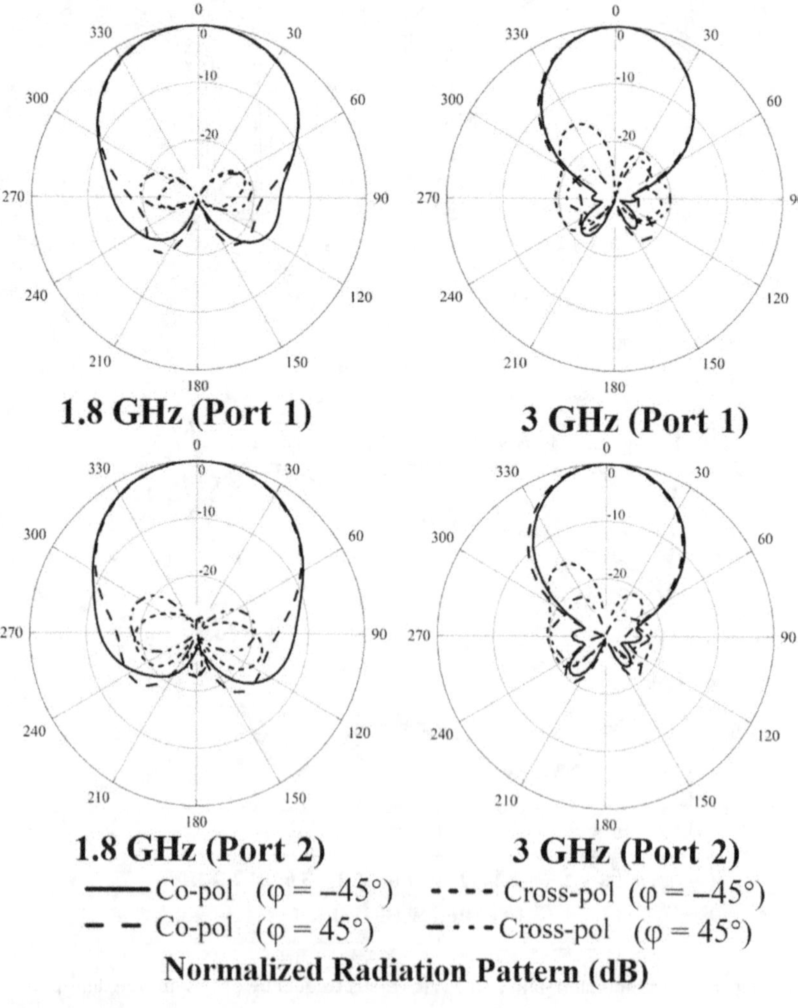

Figure 5.11 Simulated radiation patterns of port 1 and port 2 at different frequencies. Adapted from [10]

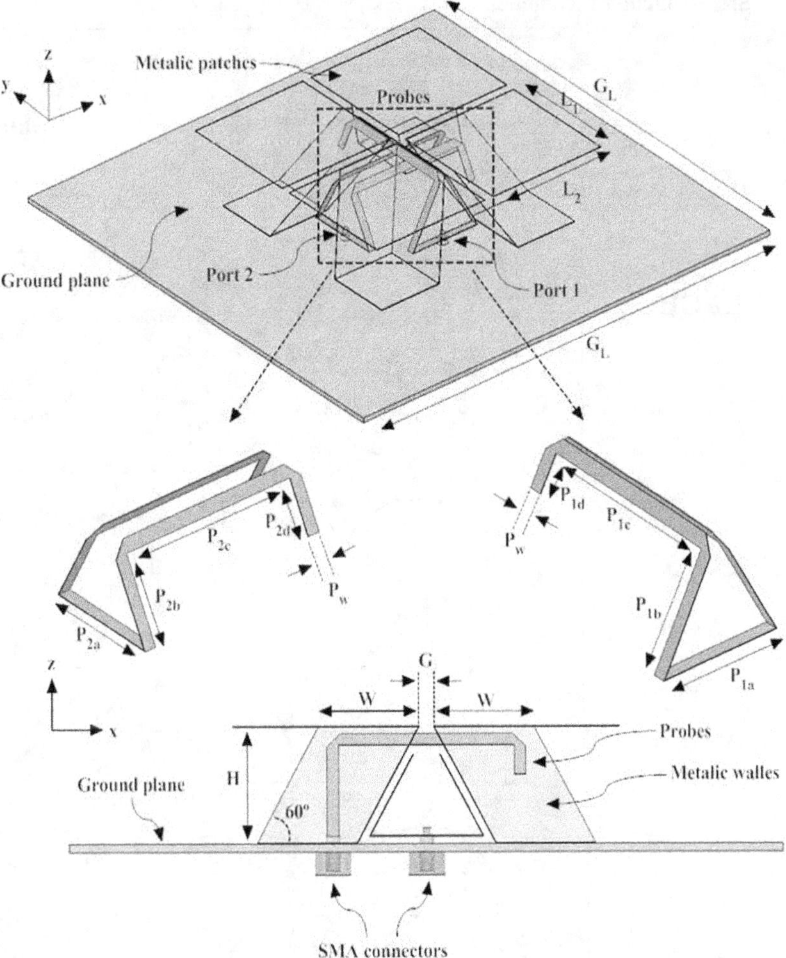

Figure 5.12 Perspective and side views of dual-polarized ME dipole antenna with dual Γ-probe feeds [11]

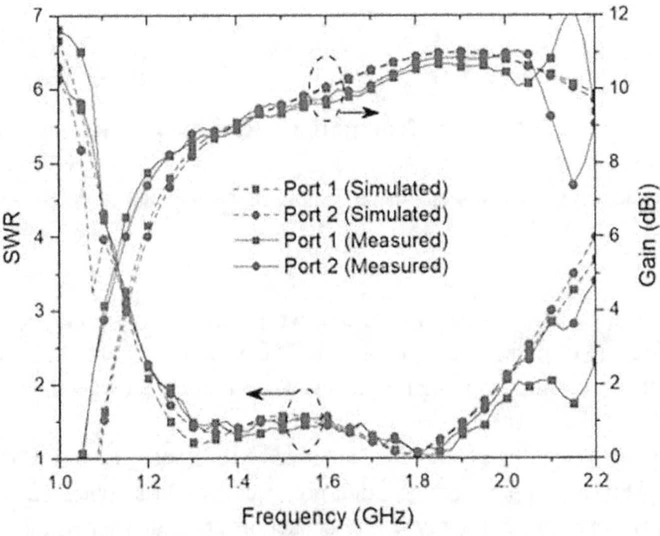

Figure 5.13 SWR and gain of dual-polarized ME dipole with dual Γ-probe feeds [11]

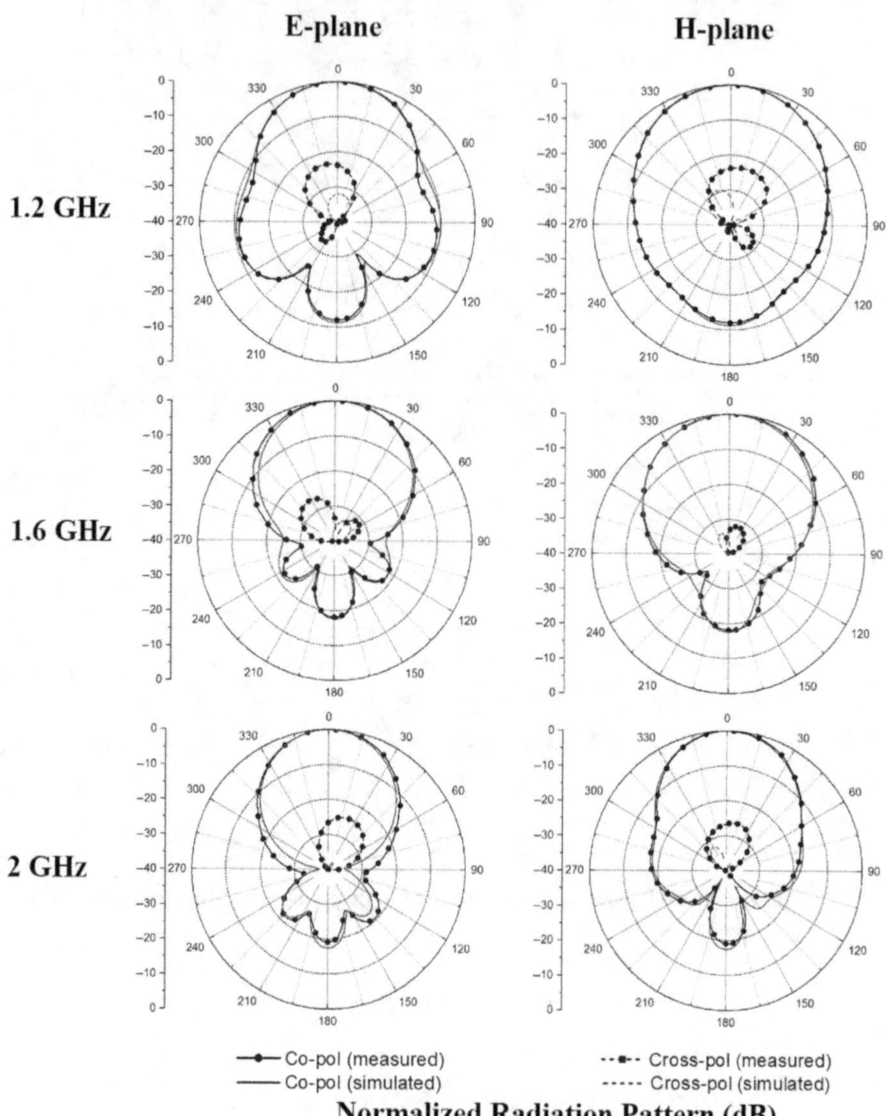

Figure 5.14 Radiation patterns of dual-polarized ME dipole with dual Γ-probe feeds measured at port 1 [11]

at port 1. As the radiation patterns of the two polarizations are almost identical, the radiation pattern of port 2 is not given here. The beamwidths of both E- and H-planes reduce with increasing operating frequency. The isolation between the two input ports is higher than 28 dB over the operating band.

The previous two low-profile dual-polarized ME dipoles are quite complicated in structure. Another simpler configuration proposed by Christophe Delaveaud's group [12] is shown in Figure 5.15, where four capacitors are added across the gaps between

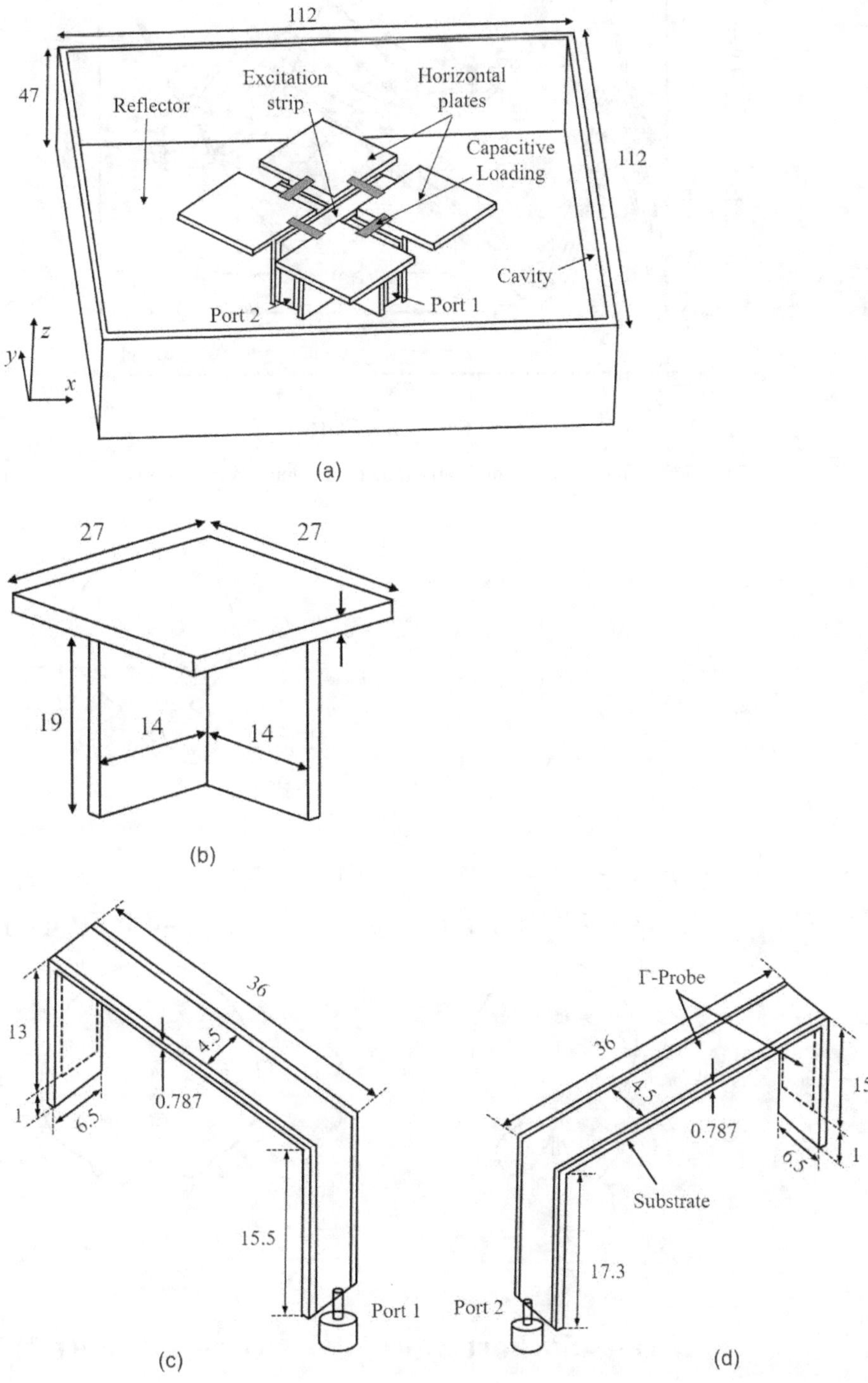

Figure 5.15 Dual-polarized ME dipole with capacitive loading. (a) Perspective view of whole structure, (b) Perspective of one quarter of the structure, and (c) Feeds for port 1 and port 2 (dimensions in mm). Adapted from [12]

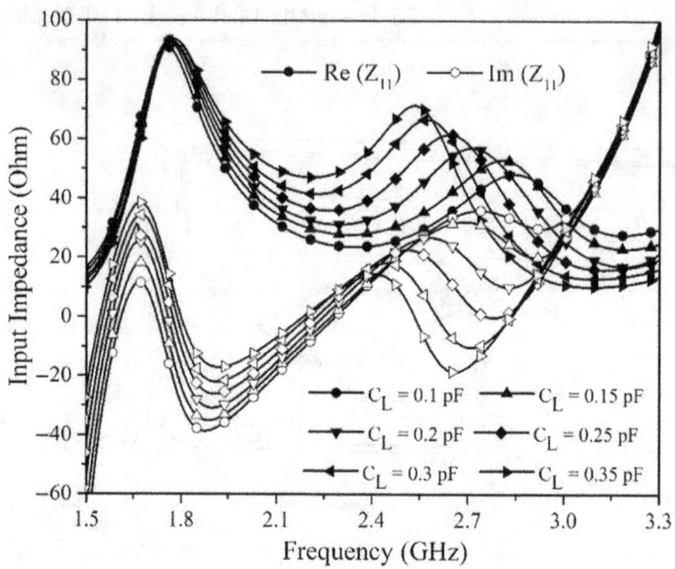

Figure 5.16 Effect of the loaded capacitors on the antenna input impedance. Adapted from [12]

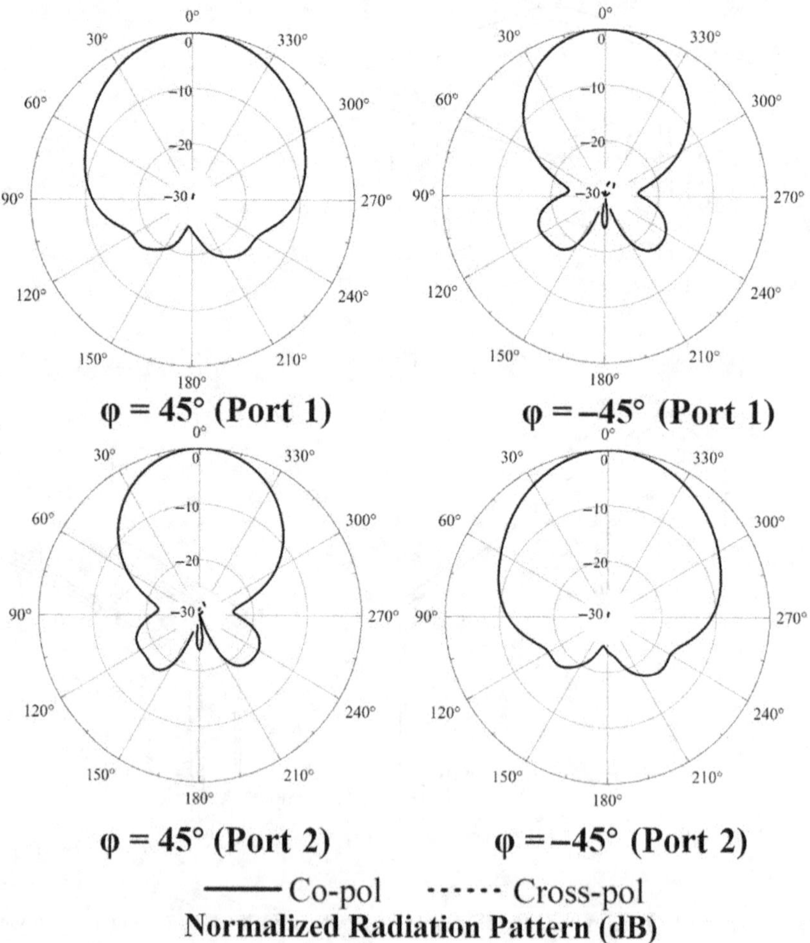

Figure 5.17 Simulated radiation patterns of dual-polarized ME dipole with capacitive loading. Adapted from [12]

the four horizontal patches of the basic design, as described in Chapter 2, with a lower height of about $0.15\lambda_0$. The resonant frequency of the magnetic dipole is shifted up when the height of the antenna is reduced. As shown in Figure 5.16, when the capacitance value is increased gradually, the resonant frequency of the quarter-wave patch antenna will be decreased to help restore the wideband performance of the ME dipole. It is observed that the capacitors have an insignificant effect on the characteristics of the electric dipole. The antenna has an impedance bandwidth of 64% (SWR < 2) and a gain increasing from 7 to 10 dBi as the operating frequency increases. The radiation pattern is similar to previous designs with a high front-to-back ratio if the side walls are selected with a higher value as shown in Figure 5.17.

5.4 ME Dipole with Small Projection Area

Antenna elements with small footprints are preferable for array designs as the elements can be placed closer together for achieving lower side lobe levels and wider scanning angles. In the design of linearly polarized ME dipoles, the horizontal electric arms can be folded to reduce the projection area, but the effect is not very substantial unless the edges of the electric dipole are shorted to the ground through some metallic pins [13], which will be described in Chapter 6 under the section on wideband and wide beamwidth designs. Here, we only describe two interesting dual-polarized ME dipoles with smaller projection area.

By replacing the four horizontal rectangular patches of a conventional dual-polarized ME dipole with four planar octagonal loops, a small dual-polarized ME dipole having 58% overlapped impedance bandwidth with SWR < 1.5 from 1.68 to 3.05 GHz was reported [14]. Stable gain of about 8.3 dBi with less than 1 dB variation and over 26 dB isolation between the two input or output ports was achieved. The antenna was placed inside a box-shaped reflector with dimensions of $0.79\lambda_0 \times 0.79\lambda_0 \times 0.26\lambda_0$, where λ_0 refers to the center operating frequency. The antenna was suggested for uses in the 2G/3G/4G base station antenna arrays.

The antenna element described earlier still has a foot print large than $0.5\lambda_0 \times 0.5\lambda_0$. For further reduction of the size, an innovative approach was proposed by Christophe Delaveaud's group [15, 16]. As depicted in Figure 5.18, the four electric dipole arms are changed from square plates to folded square loops for increasing the length of the current path in each electric dipole arm. The structure looks complex, but it was realized by a low-cost 3D printing technique. It is demonstrated that the projection area can be reduced to about $0.32\lambda_0 \times 0.32\lambda_0$, where λ_0 refers to the center operating frequency for a design exhibiting 55% bandwidth with SWR < 2.5 from 1.67 to 2.94 GHz. The antenna can maintain a high broadside gain of 8.5 dBi with 1.5 dB variation. One major weakness in this design is that the port isolation is only about 15 dB over the operating frequency band. Another weakness is that the beamwidth has a larger variation over the operating frequency band – decreasing from 76° to 60° in one port and from 70° to 60° in the other port when the operating frequency increases from 1.7 to 2.9 GHz.

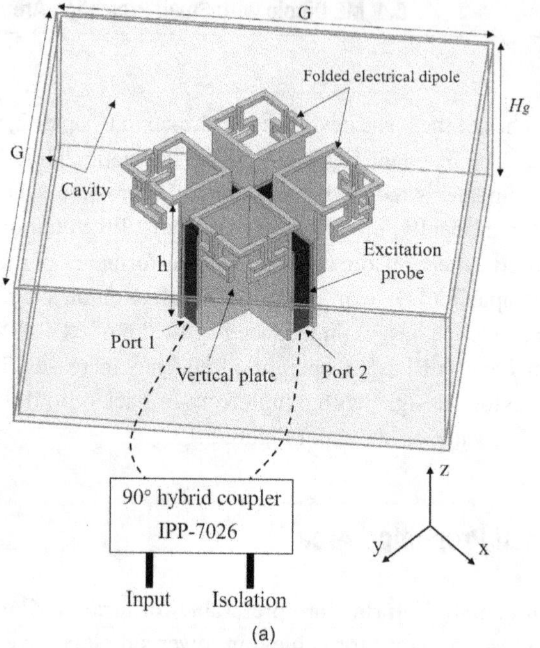

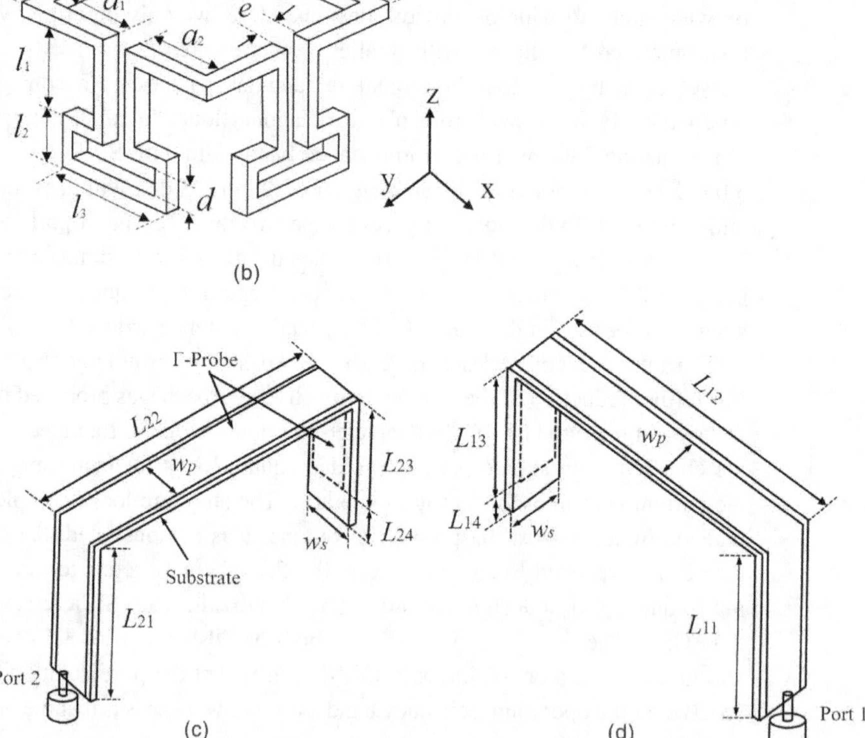

Figure 5.18 Compact dual-polarized ME dipole with small projection area. (a) Perspective view of whole structure, (b) Perspective view of one quarter of the structure, and (c) Feeds for port 1 and port 2. Adapted from [16]

5.5   Substrate-integrated Low-profile ME Dipole

Dielectric loading is a common method to reduce the size of many antennas. The first reported dielectric-loaded ME dipole is shown in Figure 5.19. In this design, the region between the two vertical parallel plates of the basic linearly polarized ME dipole is loaded with a low-loss low-dielectric constant substrate of $\varepsilon_r = 2.65$. It was demonstrated that the antenna height can be reduced to about 0.15 λ_0, referring to the center frequency of 1.9 GHz, while maintaining similar wideband performance of the basic ME dipole. After a detailed parametric study, the antenna with dimensions given in the figure caption has 48% impedance bandwidth with SWR < 2 from 2.02 to

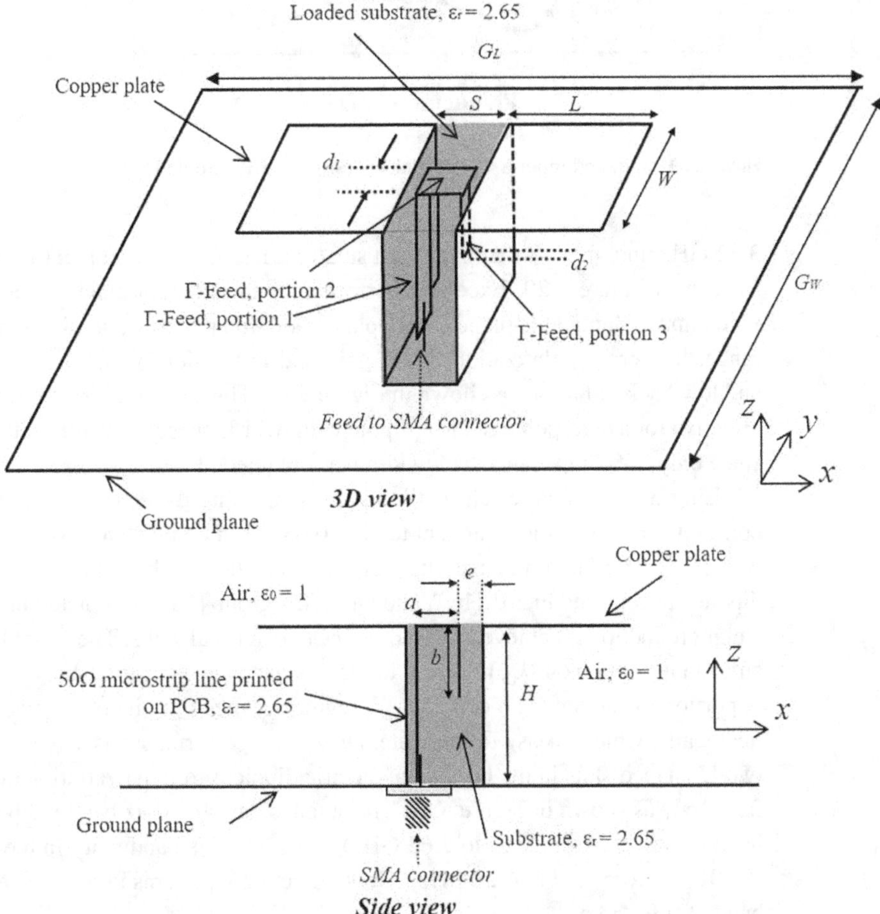

Figure 5.19 Configuration of dielectric-loaded ME dipole. (a = 7.7 mm (0.1λ_d), b = 15 mm (0.2λ_d), e = 1.5 mm (0.02λ_d), d_1 = 4 mm (0.05λ_d), d_2 = 2 mm (0.03λ_d), S = 10 mm (0.14λ_d), H = 18 mm (0.15λ_0), L = 23 mm (0.19λ_0), W = 47 mm (0.39λ_0), G_L = 150 mm (1.25λ_0), G_W = 150 mm (1.25λ_0), where λ_0 and λ_0 are wavelength in dielectric and free space, respectively, referring to the center frequency of 2.5 GHz.) [17]

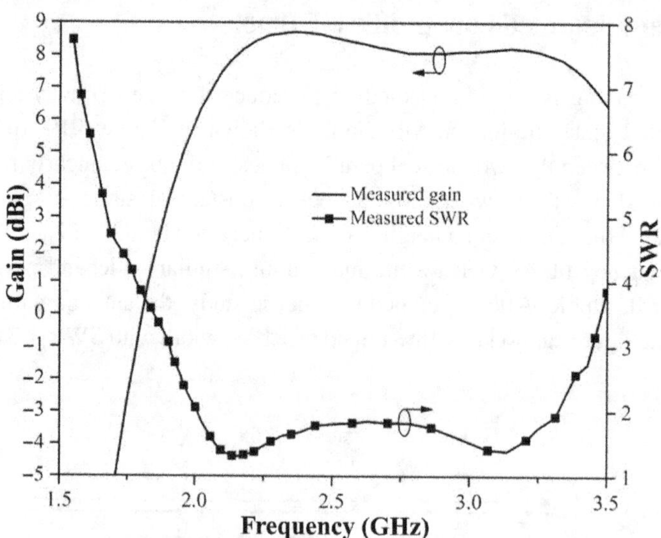

Figure 5.20 Gain and input SWR of dielectric-loaded ME dipole [17]

3.32 GHz and about 8 dBi gain with small variation over the operating frequencies, as seen in Figure 5.20. Excellent performance in radiation pattern is also achieved with almost identical E-plane and H-plane radiation patterns over the wide frequency range, low cross polarization, small gain and beamwidth variation over frequency, and low back radiation, as shown in Figure 5.21. This technique, however, is not very effective for a dual-polarized ME dipole with twin-L probe feeds for each polarization since only 25% impedance bandwidth was obtained [17].

Using nonconductive felt textile as the supporting dielectric substrate (dielectric constant $\varepsilon_r = 1.3$ and loss tangent $\tan \delta = 0.033$) and conductive textile (conductivity $\sigma \cong 1.18 \times 10^5$ S/m) as the metallic part, a low-profile dual-band linearly polarized ME dipole was reported in 2015 by Vandenbosch's group [18], as depicted in Figure 5.22. Height reduction is achieved by folding the two vertical walls. The 6 mm height of the antenna is about $0.048\lambda_0$, referring to the lower-band center frequency of 2.375 GHz. A portion of the feed is a coplanar waveguide realized in one of the two folded vertical walls, which is easy to fabricate. Dual-band performance is achieved by etching two U-shaped slots in the horizontal electric dipole. From the variation of S_{11} against frequency as shown in Figure 5.23. The antenna has 490 MHz bandwidth in the lower band (SWR < 2 from 2.13 to 2.62 GHz) and 1.71 GHz bandwidth in the upper band (SWR < 2 from 4.57 to 6.28 GHz). Good directional patterns in both lower and upper bands are found.

The on-body performance of the antenna is acceptable, as demonstrated by its low level of Specific Absorption Rate, less than 2 W/kg fulfilling the European standard. Deformation tests indicate that the antenna can still perform well under bending situation. This low-profile ME dipole is a competitive solution for on-body communications.

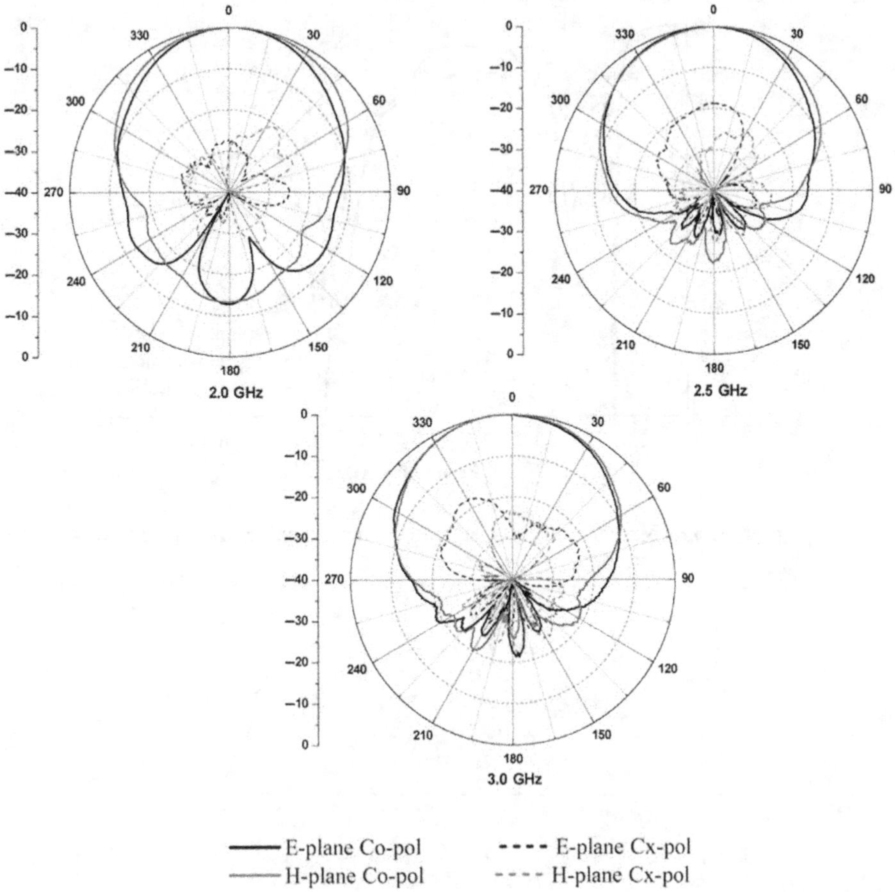

Figure 5.21 Radiation patterns of dielectric-loaded ME dipole at different frequencies [17]

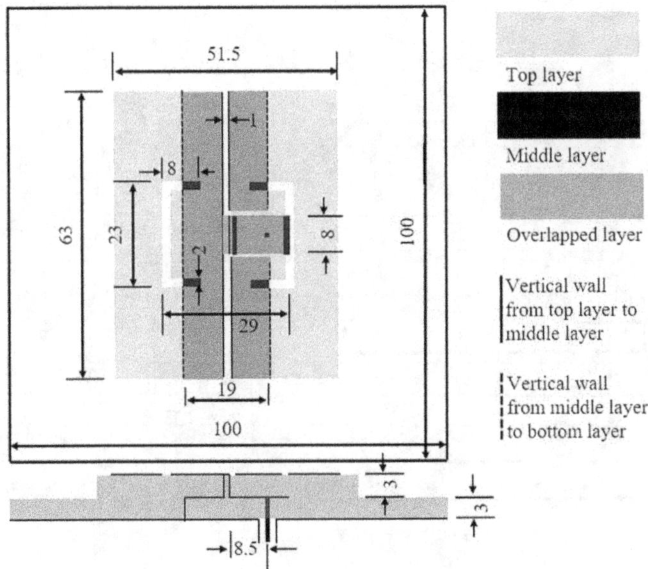

Figure 5.22 Geometry of textile ME dipole (dimensions in mm) [18]

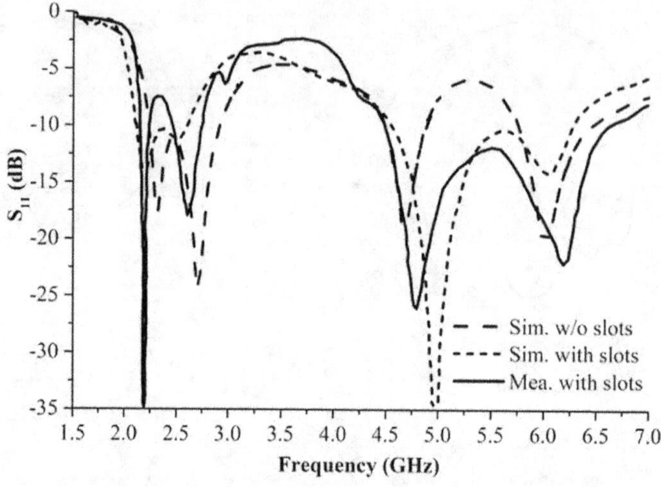

Figure 5.23 Variation of S_{11} against frequency of textile ME dipole [18]

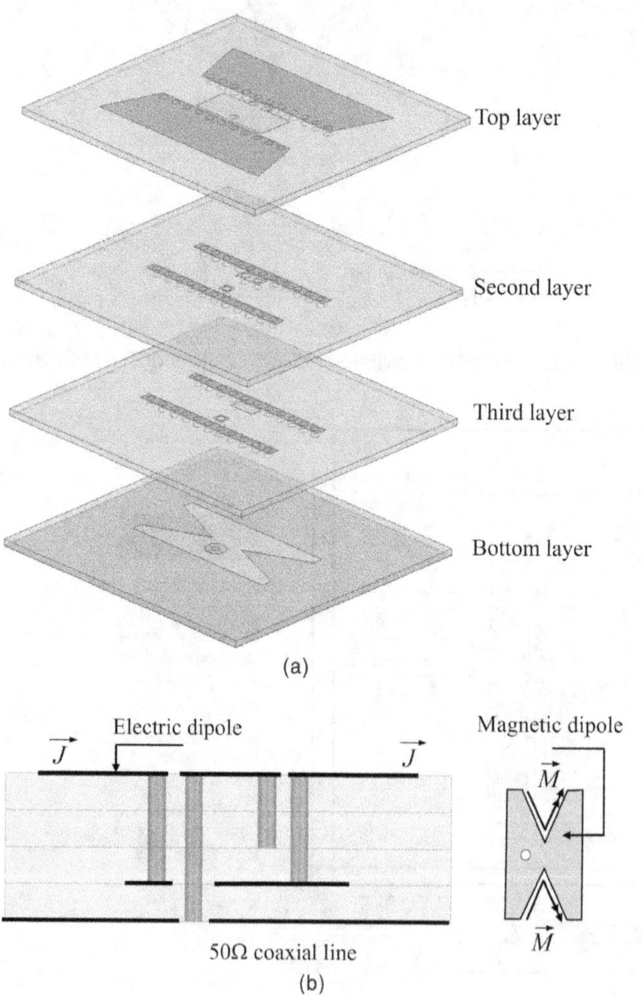

Figure 5.24 Substrate-integrated ME dipole. (a) Perspective view and (b) Side view [19]

A substrate-integrated ME dipole for 5G Wi-Fi was also reported in 2015 [19]. The antenna is printed on a stack of four dielectric substrates with dielectric constant of 2.65 and total thickness of 6 mm. As shown in Figure 5.24, the novelty in this design is that the central part of the magnetic dipole is modified to become a tapered H-shaped plate. Unlike other designs, this part is separated from the grounded plane by a dielectric layer. This approach can reduce the resonant frequency of the shorted patch antenna as the length of the current path in the antenna is increased. Using PCB fabrication technique, the vertical walls can only be realized with a number of shorting pins. Without the use of folding walls, a low-profile ME dipole with thickness of $0.11\lambda_0$ was achieved.

Although the experimental results are not as good as the simulation results due to fabrication tolerance, good performance of the antenna was confirmed. Experimentally, the antenna has an impedance bandwidth of about 20% with SWR < 2 from about 5 to 6 GHz, and the gain varies between 6 and 7 dBi over this frequency range, as shown in Figure 5.25. As the resonant frequencies of the electric dipole and magnetic dipole are not different too much, the complementary effect is very effective. It can be seen in Figure 5.26 that the radiation patterns over the operating frequencies are of cardiac shape in both E- and H-planes. More than 20 dB front-to-back ratio can be observed. The beamwidth variation is less than 10° in both principal planes over the 20% impedance bandwidth. It should be mentioned that the width of the horizontal portion of the Γ-shaped feed is increased to improve the impedance matching of the antenna.

Further enhancement of the above design was reported in 2020 by another group [20]. Incorporating the folding wall technique similar to that shown in Figure 5.22,

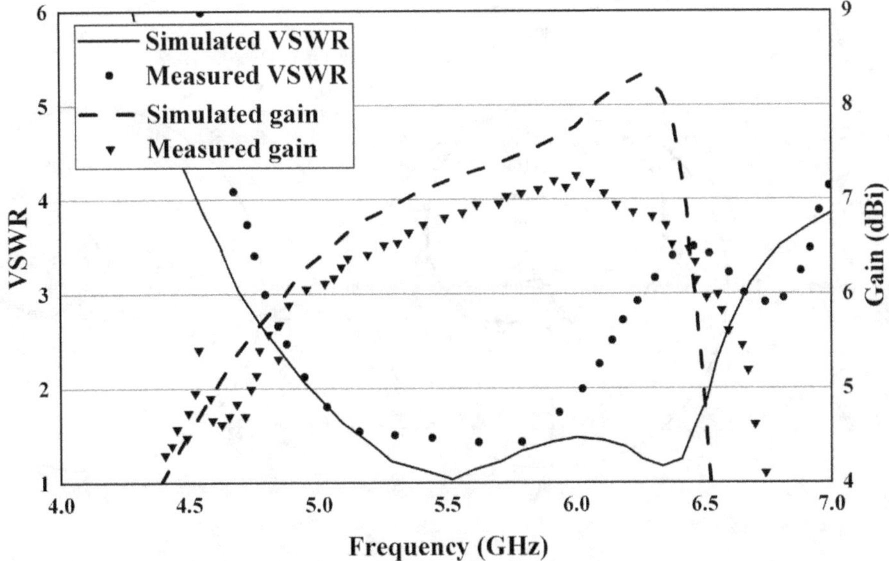

Figure 5.25 Input SWR and gain of substrate-integrated ME dipole [19]

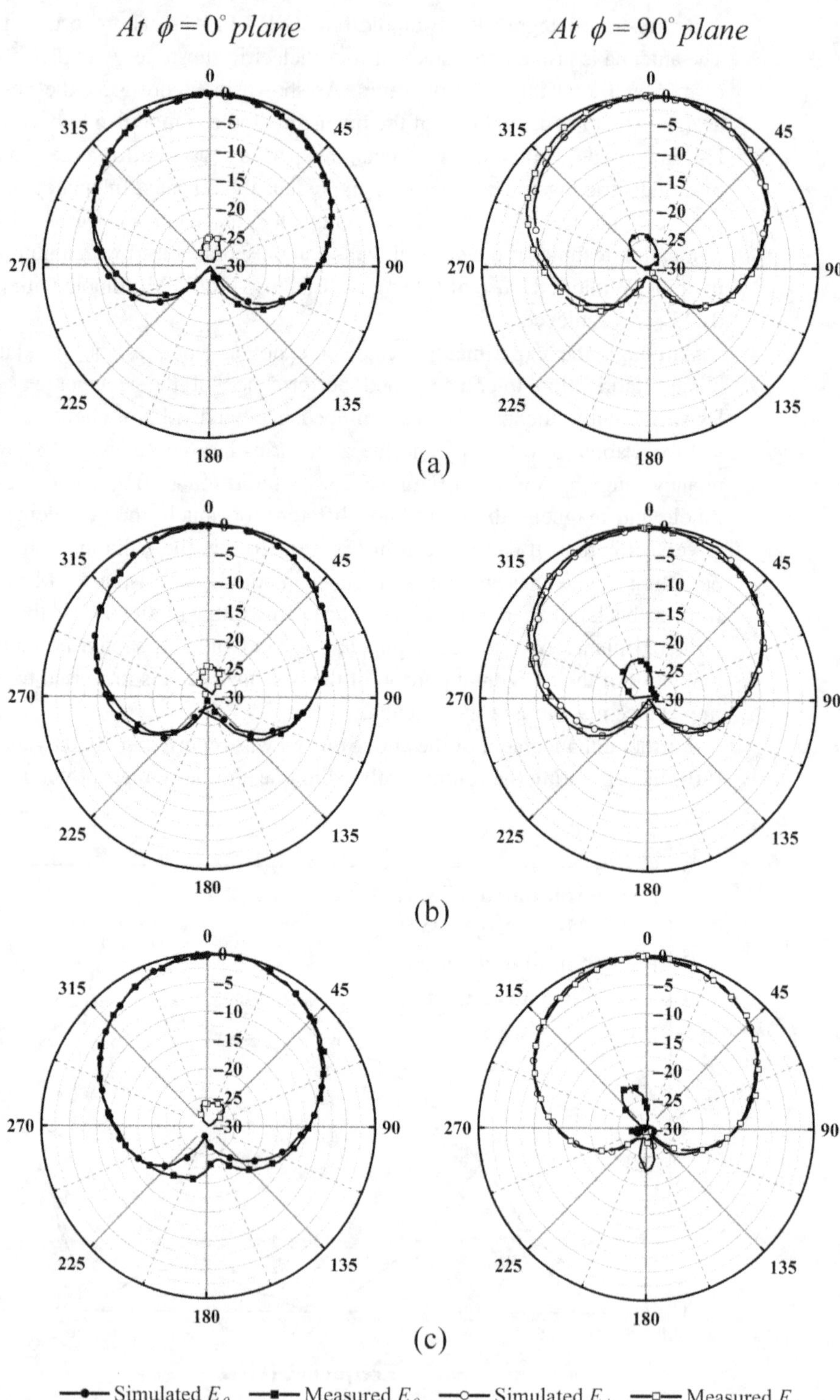

Figure 5.26 Radiation patterns of substrate-integrated ME dipole [19]. (a) 5.1 GHz, (b) 5.5 GHz, and (c) 5.8 GHz

a substrate-integrated folded ME dipole antenna printed on a three-layered dielectric substrate with total thickness of $0.12\lambda_0$ was demonstrated with more than 34% impedance bandwidth and stable gain varying between 6 and 7.5 dBi over the operating frequencies.

Another low-profile substrate-integrated ME dipole antenna proposed for UWB Indoor Positioning Systems by Prof. Xiuping Li's group is shown in Figure 5.27 [20]. By folding each vertical wall four times as the design depicted in Figure 5.1 and using FR-4 substrates with dielectric constant $\varepsilon_r \cong 4.4$, the antenna can perform with 20% impedance bandwidth (SWR < 2 from 3.73 to 4.55 GHz) for a small thickness of $0.064\lambda_0$. Like other ME dipoles, the antenna exhibits a cardiac-shaped radiation pattern with high front-to-back ratio. The gain of the antenna is around 6 dBi with small variation over the operating frequencies. It is worth to mention that the feed of this antenna is connected to a grounded coplanar waveguide for ease of integration with MMIC. Similar folded structure was also reported [21]

In the design of a substrate-integrated aperture-coupled dual-polarized ME dipole antenna with filtering function, an interesting technique was proposed to reduce the

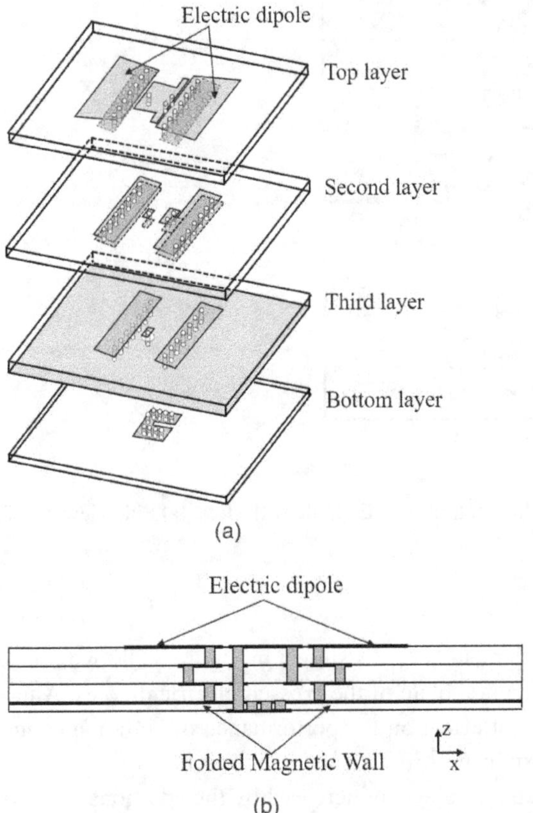

Figure 5.27 Substrate-integrated ME dipole connecting to GCPW line. (a) Disassembled perspective view and (b) Side view [20]

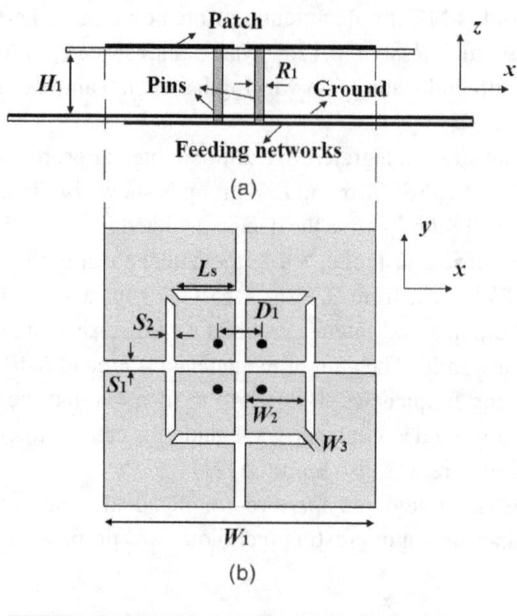

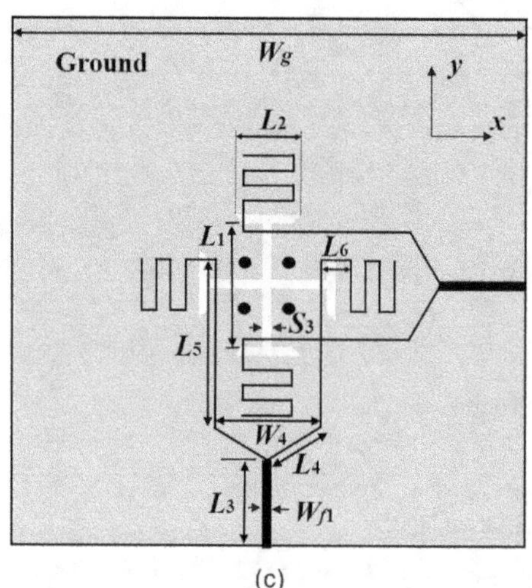

Figure 5.28 Geometry of dual-polarized ME dipole with slots. (a) Side view, (b) Top view, and (c) Bottom view [22]

height of the antenna element [22]. As shown in Figure 5.28, two open-ended slots are etched in each horizontal plate of the crossed electric dipoles. Amazingly, their presence has a significant effect on the performance of both magnetic and electric dipoles. A possible explanation is that the different paths of the current flow for the electric and magnetic dipoles are increased by the additional slots, so a smaller thickness of the antenna can be selected to maintain the required resonant frequency for the magnetic dipole, while smaller horizontal plates can be used to maintain the

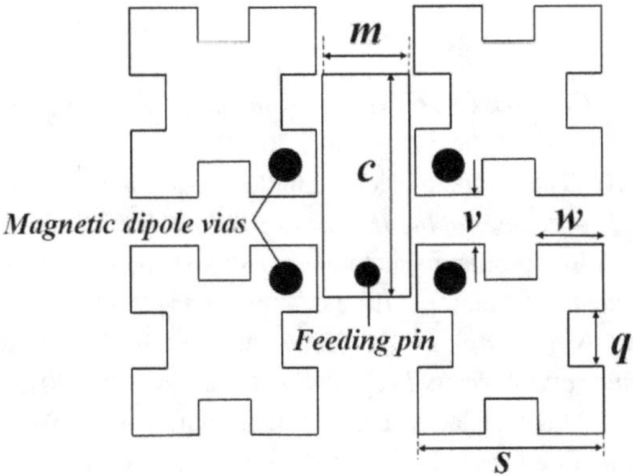

Figure 5.29 Geometry of an ME dipole with fractal structure ($m = 2.7$ mm, $c = 6$ mm, $s = 5.2$ mm, $v = 1.3$ mm, $w = 1.5$ mm, $q = 1.3$ mm). Adapted from [23]

required resonant frequency for the electric dipole. For the reported prototype with 27.6% impedance bandwidth with SWR < 2 between 3.3 and 4.2 GHz, the height is only $0.11\lambda_0$ and the projection area is $0.45\lambda_0 \times 0.45\lambda_0$.

In the design of a wideband ME dipole with low radar cross section (RCS) by Prof. Xiangyu Cao's group [23], a technique for reducing the projection area of the ME dipole was proposed. As depicted in Figure 5.29, four horizontal Minkowski-like fractal metallic plates are used to realize the electric dipole, which is printed on a 4 mm-thick dielectric substrate. The magnetic dipole is realized by four vertical plate-through-holes connecting the four plates to the grounded plane. The antenna is excited by an L-shaped feed. With the geometric parameters given in the caption of Figure 5.29, the antenna exhibits 42.4% impedance bandwidth with SWR < 2 from 8.0 to 12.3 GHz. The projection area is only around $0.45\lambda_0 \times 045\lambda_0$. Owing to its low radiation level in the horizontal directions, this ME dipole antenna can maintain its excellent characteristics when it is incorporated with artificial magnetic conducting structures printed on the same dielectric substrate for reducing the RCS substantially over 80% bandwidth. Details can be referred to in the original paper [23].

5.6 Summary

Major techniques for reducing the height and projection area of the ME dipole available in the literature are reviewed, including those by the author's group and other research groups globally. The works by the author's group are reported in more detail. On the other hand, although the works by other's groups are briefly reviewed, we have simulated some of their designs to help us understand more about the physics behind. It is believed that some of these techniques can be combined together to generate more compact designs. The use of other natural materials or metamaterials may be able to create more compact designs in the future.

References

[1] K. M. Luk and H. Wong, *Complementary Wideband antenna*, US Patent, No.: US7,843,389 B2, Filed: March 10, 2006, Published: November 30, 2010.

[2] K. M. Luk and B. Q. Wu, The magneto-electric dipole – A new antenna element for wireless communications, *Proceedings of the IEEE*, vol. 100, 2012, no. 7, pp. 2297–2307.

[3] L. Ge and K. M. Luk, A low-profile magneto-electric dipole antenna, *IEEE Transactions on Antennas and Propagation*, vol. 60, 2012, no. 4, pp. 1684–1689.

[4] L. Ge and K. M. Luk, A magneto-electric dipole antenna with low-profile and simple structure, *IEEE Antennas and Wireless Propagation Letters*, vol. 12, 2013, pp. 140–142.

[5] L. Ge, S. Gao and M. Li, Magnetoelectric dipole antenna with low profile, *IEEE Antennas and Wireless Propagation Letters*, vol. 17, 2018, no. 10, pp. 1760–1763.

[6] C. Ding and K. M. Luk, Low-profile magneto-electric dipole antenna, *IEEE Antennas and Wireless Propagation Letters*, vol. 15, 2016, pp. 1642–1644.

[7] M. Li, K. M. Luk, L. Ge and K. Zhang, Miniaturization of magnetoelectric dipole antenna by using metamaterial loading, *IEEE Transactions on Antennas and Propagation*, vol. 64, 2016, no. 11, pp. 4914–4918.

[8] B. Feng, K. L. Chung, J. Lai and Q. Zeng, A conformal magneto-electric dipole antenna with wide H-plane and band-notch radiation characteristics for sub-6-GHz 5G base station, *IEEE Access*, vol. 7, 2019, pp. 17470–17479.

[9] M. Li and K. M. Luk, A low-profile wideband planar antenna, *IEEE Transactions on Antennas and Propagation*, vol. 61, 2013, no. 9, pp. 4411–4418.

[10] A. S. Kaddour, S. Bories, A. Bellion and C. Delaveaud, Low profile dual-polarized wideband antenna, *Proceedings of ISAP2016*, Okinawa, pp. 86–87.

[11] C. Ding and K. M. Luk, A low-profile dual-polarized magneto-electric dipole antenna, *IEEE Access*, vol. 7, 2019, pp. 181924–181932.

[12] A. S. Kaddour, S. Bories, C. Delaveaud and A. Bellion, Wideband dual-polarized magneto-electric miniaturization using capacitive loading, *IEEE APS International Symposium Digest*, 2017, pp. 545–546.

[13] G. Yang, J. Li, J. Yang and S. G. Zhou, A wide beamwidth and wideband magnetoelectric dipole antenna, *IEEE Transactions on Antennas and Propagation*, vol. 66, 2018, no. 12, pp. 6724–6733.

[14] Z. Li, Y. Sun, M. Yang, P. Tang and Z. Wu, A compact dual-polarized magneto-electric dipole antenna for 2G/3G/LTE applications, *Proceedings of 2017 Progress in Electromagnetics Research Symposium*, Singapore, November 2017, pp. 1338–1344.

[15] A. S. Kaddour, S. Bories, A. Bellion and C. Delaveaud, 3D printed compact dual-polarized wideband antenna, *Proceedings of 2017 11th European Conference on Antennas and Propagation*, 2017, pp. 3441–3443.

[16] A. S. Kaddour, S. Bories, A. Bellion and C. Delaveaud, 3-D-printed compact wideband magnetoelectric dipoles with circular polarization, *IEEE Antennas and Wireless Propagation Letters*, vol. 17, 2018, no. 11, pp. 2026–2030.

[17] L. Siu, H. Wong and K. M. Luk, A dual-polarized magneto-electric dipole with dielectric loading, *IEEE Transactions on Antennas and Propagation*, vol. 57, 2009, no. 3, pp. 616–623.

[18] S. Yan, P. J. Soh and G. A. E. Vandenbosch, Wearable dual-band magneto-electric dipole antenna for WBAN/WLAN applications, *IEEE Transactions on Antennas and Propagation*, vol. 63, 2015, no. 9, pp. 4165–4169.

[19] H. W. Lai and H. Wong, Substrate integrated magneto-electric dipole antennas for 5G Wi-Fi, *IEEE Transactions on Antennas and Propagation*, vol. 63, 2015, no. 2, pp. 870–874.

[20] X. Li, Q. Li, H. Zhu, J. Song and Z. Qi, A low-profile substrate integrated magneto-electric dipole antenna based on folded magnetic wall for UWB application, *IEEE MTT-S International Microwave Symposium Digest*, 2018, pp. 1545–1548.

[21] Z. Du and Q. Feng, A wideband folded magnetoelectric dipole antenna with miniaturized integration for 5G applications, *Journal of Electromagnetic Waves and Applications*, vol. 34, 2020, no. 12, pp. 1631–1646.

[22] S. J. Yang, Y. M. Pan, Y. Zhang, Y. Gao and X. Y. Zhang, Low-profile dual-polarized filtering magneto-electric dipole antenna for 5G applications, *IEEE Transactions on Antennas and Propagation*, vol. 67, 2019, no. 10, pp. 6235–6243.

[23] C. Zhang, X. Y. Cao, J. Gao, S. J. Li and Y. J. Zheng, Low RCS and broadband ME dipole antenna loading artificial magnetic conductor structures, *Radio Engineering*, vol. 26, 2017, no. 1, pp. 38–44.

6 Bandwidth Enhancement Techniques for Magnetoelectric Dipoles

6.1 Introduction

Antennas with wide bandwidth are preferable in many practical applications. The obvious reason is that the installation space can be substantially reduced if only one antenna can be employed to serve several wireless channels or even multiple wireless or mobile systems. This is a challenging task, as different wireless systems may require different radiation patterns or polarizations. It is also difficult to achieve if different wireless channels need the same radiation pattern, gain, and beamwidth. In parallel, antennas with exceedingly wide bandwidth are also required to support the development of the ultra-wideband (UWB) communication system and radar systems.

In this chapter, major techniques for enhancing the bandwidth of magnetoelectric (ME) dipoles available in the literature are reviewed and discussed. Designs with single input port and differential input ports are reported. Hopefully, it can help the readers to appreciate the beauty of these interesting designs and inspire innovative designs for future applications.

6.2 Wideband Linearly Polarized ME Dipoles

The first approach to enhance the bandwidth of the ME dipole is to employ a pair of Γ-shaped feeds as described in Chapter 3. About 70% impedance bandwidth (with SWR < 2 from 1.53 to 3.17 GHz) was achieved. This gave the motivation to develop other methods as the twin-feed structure is quite complicated in fabrication. Among the early approaches developed for enhancing the impedance bandwidth of the ME dipole, the one with two E-shaped patches is worthy to understand [1]. A design example of this antenna is shown in Figure 6.1. Each arm of the electric dipole is basically in the E-shaped form. It can be observed that the central part of each E-shaped patch has a trapezoidal shape to improve the matching performance. Also, the usual capacitive-coupling feed is modified with the ended portion going outside the two vertical parallel walls and folded into an L-shape. With these two changes, the ME dipole can perform with about 60% impedance bandwidth (SWR < 1.5). The gain and radiation pattern remain stable over the operating frequency range from 1.84 to 3.44

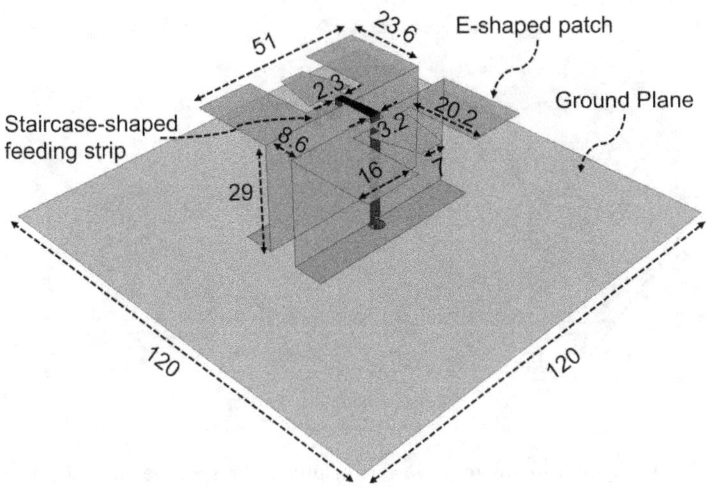

Figure 6.1 ME dipole with E-shaped patches. Adapted from [1]

Figure 6.2 ME dipole with C-shaped patches. Adapted from [2]

GHz. The average gain is about 8 dBi and the cross polarization is less than −20 dB over the operating frequencies.

Another similar design is depicted in Figure 6.2, where the electric dipole is realized in the form of two back-to-back C-shaped patches placed horizontally [2]. The antenna is excited by the common Γ-shaped feed. With appropriate dimensions, the antenna exhibits 60% impedance bandwidth with SWR < 2 between 1.64 and 3.04 GHz. Using a box-shaped reflector, more than a 20 dB front-to-back ratio and an average gain of 8.9 dBi are achieved over the operating frequency range.

By folding the two vertical walls of the ME dipole together with the first portion of the feedline connecting to the connector as shown in Figure 6.3, this not only can effectively reduce the height of the antenna but also can increase the bandwidth if the feedline is modified to have a tapered width and a tuning patch is attached to the horizontal portion of the feedline. With appropriate dimensions [3], the antenna

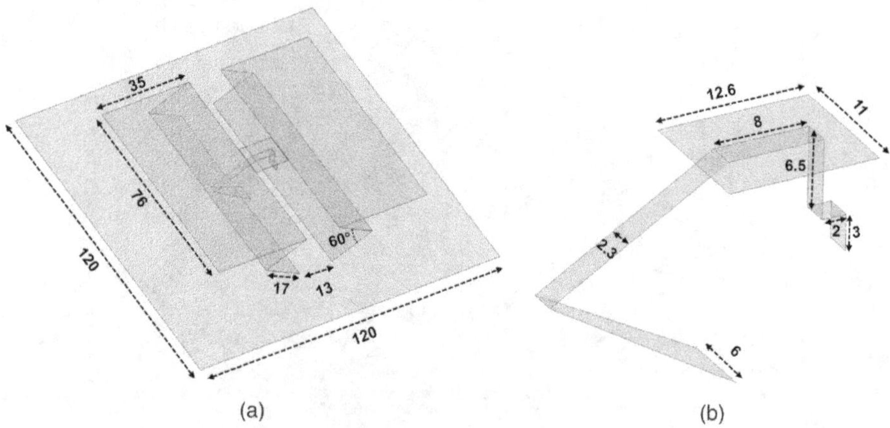

Figure 6.3 ME dipole with improved feed structure. (a) Perspective view of antenna and (b) Perspective view of feed. Adapted from [3]

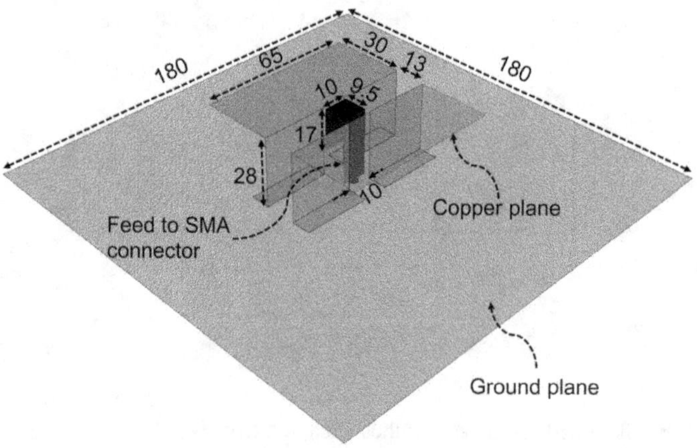

Figure 6.4 ME dipole with defected-ground microstrip line feed. Adapted from [4]

shows 86% impedance bandwidth with SWR < 2 from 1.57 to 3.95 GHz. Although the bandwidth is better than the previous designs, the gain, beamwidth, and radiation pattern have larger variation over the operating frequencies, particularly at upper frequencies. The gain decreases gradually from 9.7 to 6.2 dBi when the operating frequency increases from 1.7 to 3.9 GHz. The beamwidth is wider at the upper frequencies.

The separation between the electric dipole resonance and the magnetic dipole resonance of the ME dipole can be used to control the bandwidth of the antenna. To achieve wider bandwidth, the separation should be increased. However, if the separation is too large, the antenna will become a dual-band structure.

It is found that by simply removing the portions of the sidewalls behind the Γ-shaped feedline, as shown in Figure 6.4, good impedance matching can be obtained over a

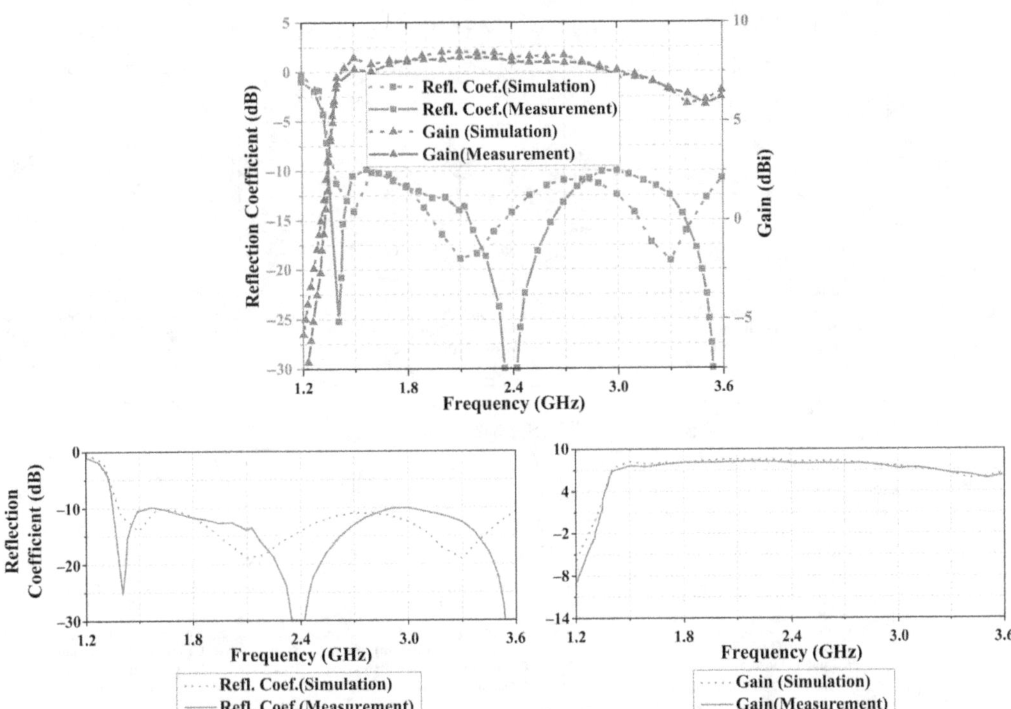

Figure 6.5 SWR and gain of ME dipole with defected-ground line feed. Adapted from [4]

wider frequency range compared with the original design. In the design [4], although the electric dipole resonance and magnetic dipole resonance are further apart compared with the original design, an overall wideband behavior can be achieved due to the use of the microstrip feedline with a defected grounded plane. As depicted in Figure 6.5, the antenna exhibits 87% impedance bandwidth with SWR < 2 from 1.38 to 3.5 GHz. The gain is around 8 dBi over the lower and central operating frequencies and drops to about 6 dBi at upper operating frequencies. The back radiation is below −20 dB over a wide frequency range except near the lowest operating frequency, as shown in Figure 6.6.

The previous design has large gain variation, although it is very wide in bandwidth. For applications requiring wideband flat gain, the antenna structure, as shown in Figure 6.7, can be considered [5]. In this geometry, a substrate-supported printed director is added on top of the electric dipole to help increase the gain at higher operating frequencies. In addition, a horn-like reflector is used to reduce the back radiation. It is interesting that although the planar electric dipole is narrower in width compared with traditional designs, the antenna has 8.1 dBi in gain with ±0.2 dB variation ranging from 1.6 to 3.8 GHz, over which SWR < 2. The disadvantage of this design is the increase in the height of the whole structure.

For applications allowing large projection area of the antenna, the one presented in Figure 6.8 may be considered if wider bandwidth is required [6]. It can be observed that the antenna has a bowtie antenna as the electric dipole. The two vertical side walls are modified into a trapezoidal shape to maintain wideband performance, which are

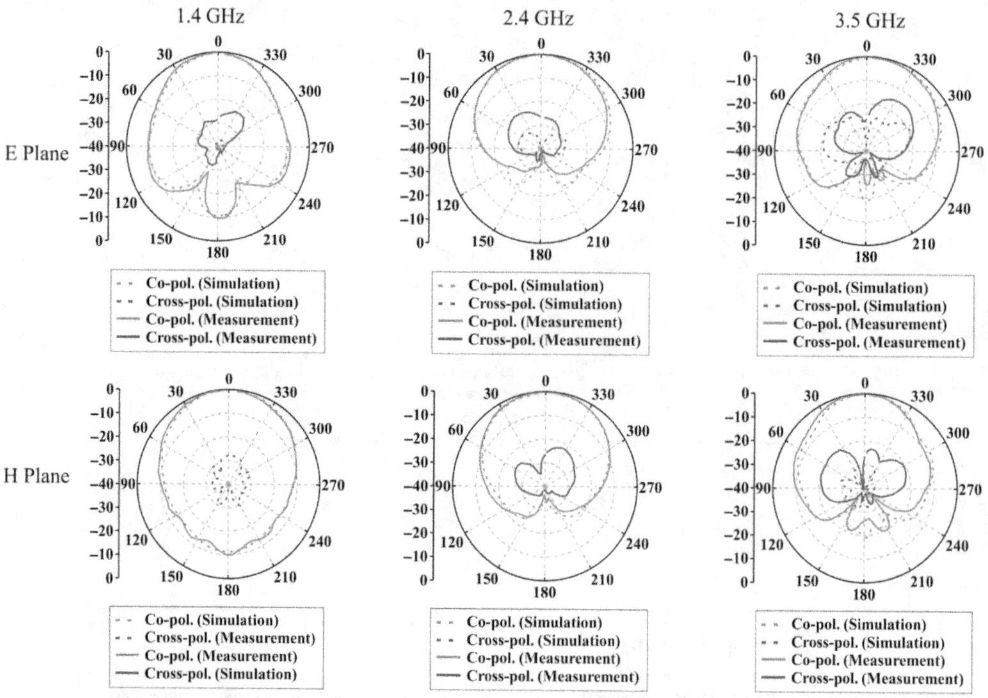

Figure 6.6 Radiation pattern of ME dipole with defected grounded line feed. Adapted from [4]

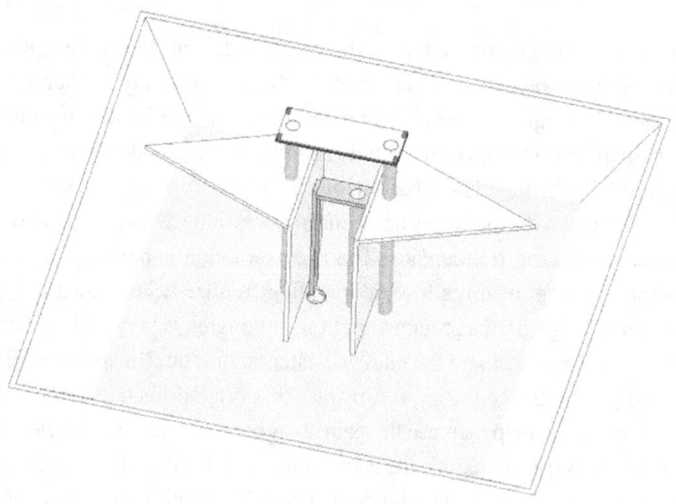

Figure 6.7 ME dipole with director for flat gain performance. Adapted from [5]

folded two times to achieve a low-profile structure. The feed line is also folded several times to make it conformal to the folded plates for realizing the magnetic dipole. The impedance bandwidth of this antenna is 95% with SWR < 2 from 1.65 to 4.65 GHz, as shown in Figure 6.9. Stable radiation pattern, as depicted in Figure 6.10, and stable gain of 7.9 ± 0.9 dBi are demonstrated.

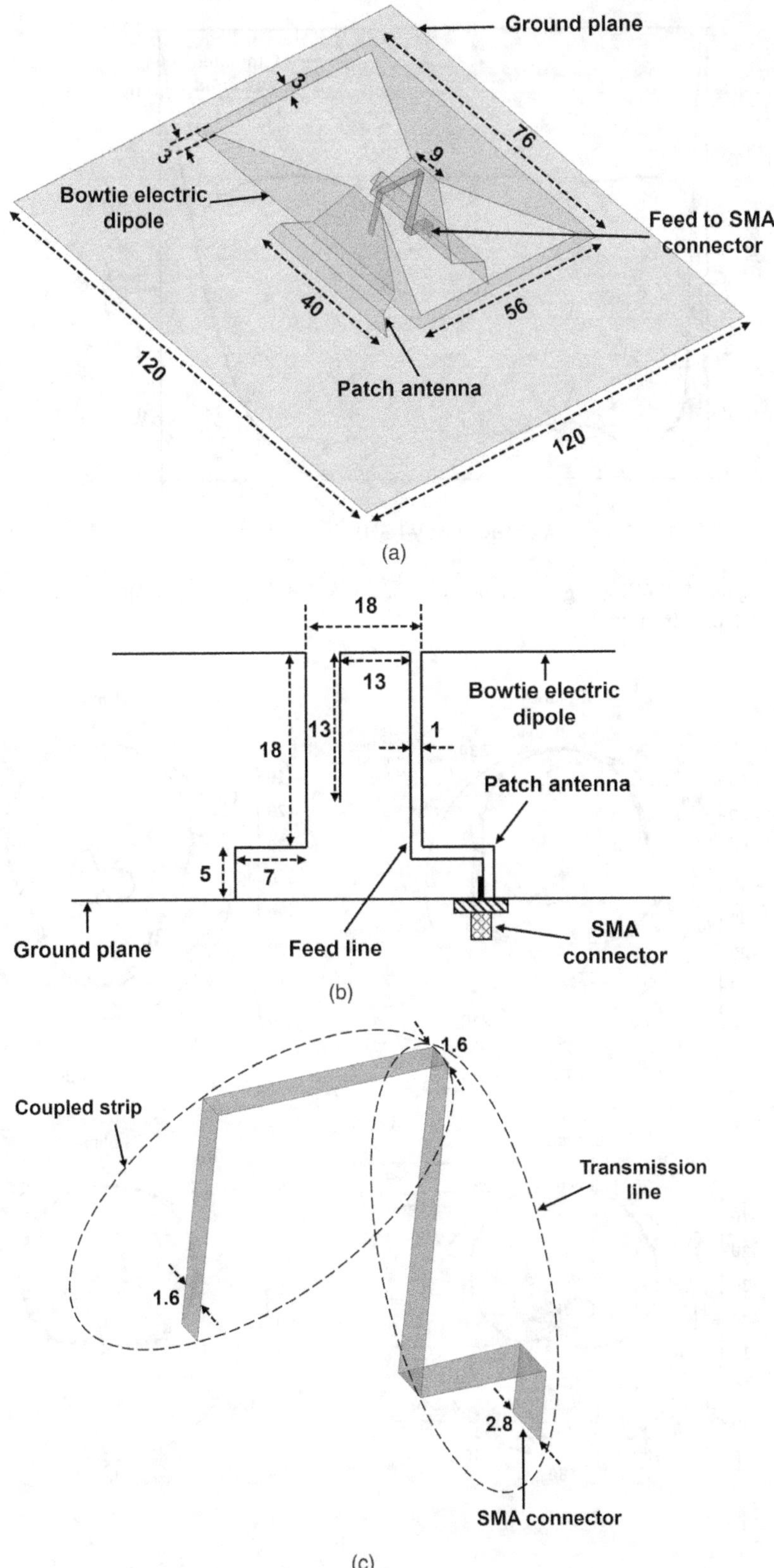

Figure 6.8 A wideband ME dipole antenna with bowtie patches. (a) Perspective view, (b) Side view, and (c) Feed structure. Adapted from [6]

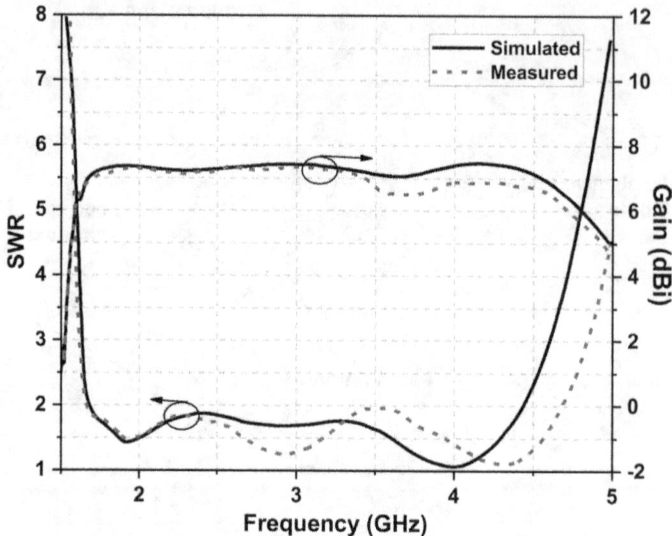

Figure 6.9 SWR and gain versus frequency of ME dipole antenna with bowtie patches. Adapted from [6]

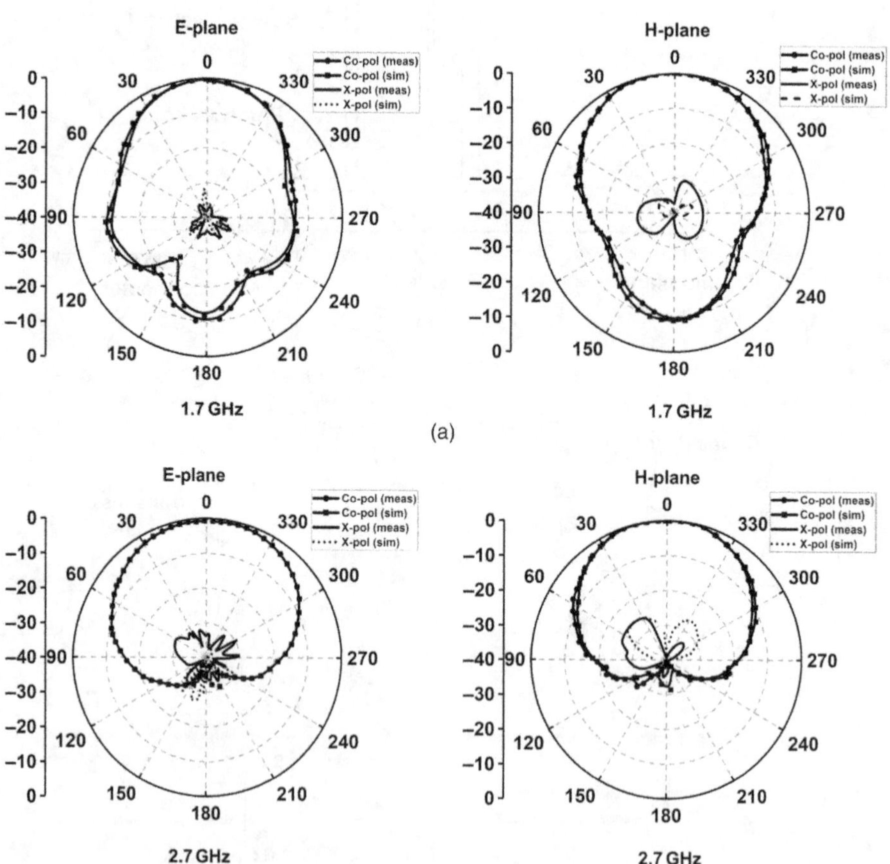

Figure 6.10 Radiation patterns of ME dipole with bowtie patches at different frequencies. Adapted from [6]

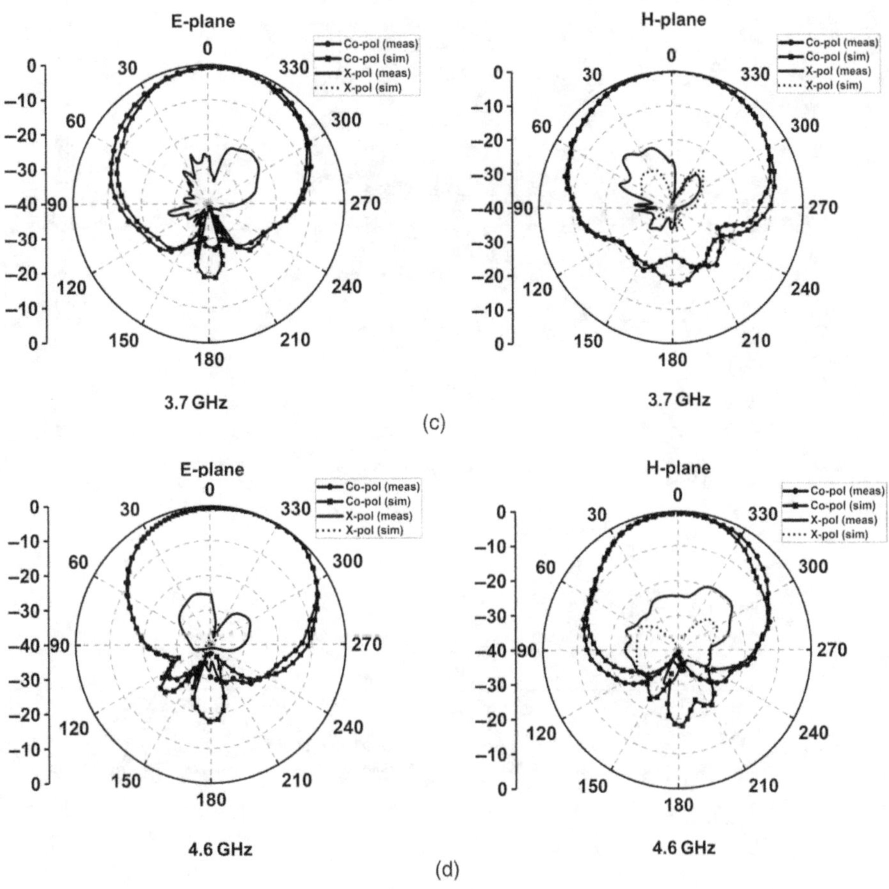

Figure 6.10 (cont.)

6.3   Wideband Dual-polarized ME Dipoles

Dual-polarized antennas are popularly used in mobile communications for enhancing channel capacity through combating multipath fading by the polarization diversity technique. There are two approaches proposed in the literature to increase the bandwidth of dual-polarized ME dipole antennas. The first method is to use a double-layered planar electric dipole. Two typical examples are shown in Figure 6.11 (version 1) [7] and Figure 6.12 (version 2) [8]. The basic concept of these two designs is the same, as the bandwidth of the electric dipole mode is extended by using two layers of horizontal patches which are connected together in the inner end points through vertical metallic posts. Although the shapes of the horizontal patches are not the same, the major difference is in the feed locations. The one with the lower patches of pentagonal shape having the vertical portions of the Γ-shaped feeds located outside the central region [7] has an impedance bandwidth of 57% with SWR < 2 from 1.5 to 2.7 GHz and more than 30 dB port

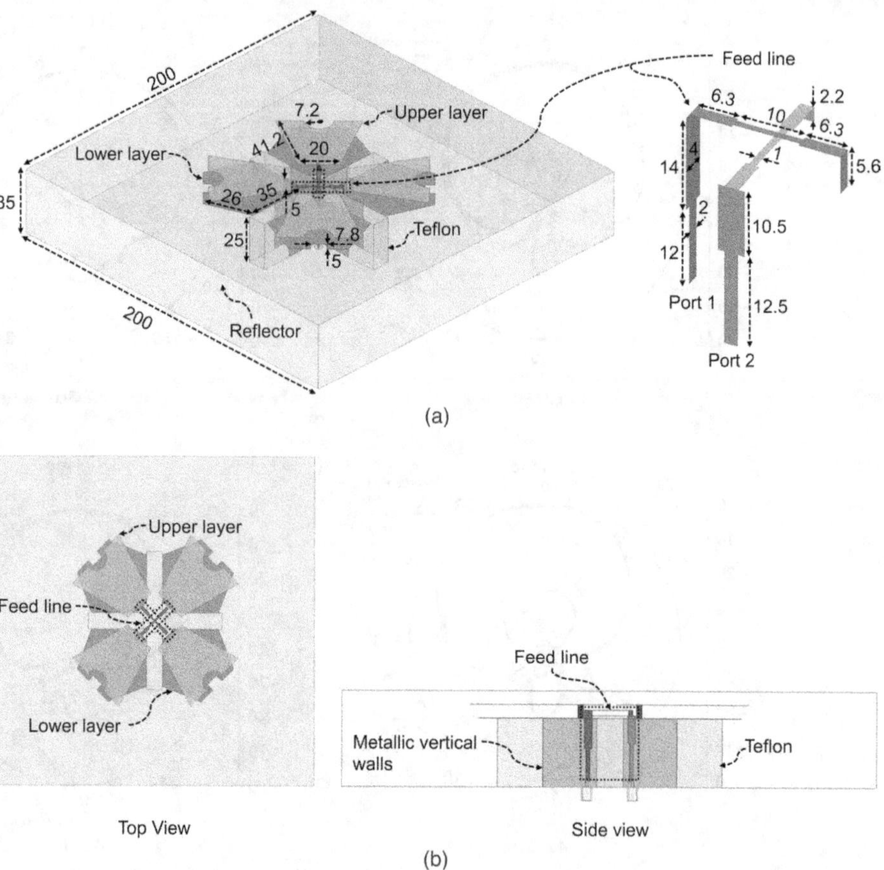

Figure 6.11 Wideband dual-polarized ME dipole with stacked electric dipoles – version 1. (a) Perspective view and (b) Top view and side view. Adapted from [7]

isolation, whereas the second one with bowtie-shaped patches having the whole Γ-shaped feed located inside the central region [8] has over 90% impedance bandwidth with SWR < 2 from 2.9 to 7.8 GHz and only 15 dB port isolation over the operating frequencies. Both structures have a box-shaped reflector to improve its overall performance.

Another approach for enhancing the bandwidth of the dual-polarized ME dipole is shown in Figure 6.13 [9]. In this design, each of the vertical walls forming the magnetic dipole mode is folded two times and the feed lines are also folded accordingly. It allows more degrees of freedom in designing the feed lines to achieve wider bandwidth. The prototype exhibits 80% impedance bandwidth with SWR < 1.5 from 1.68 to 3.02 GHz and higher than 25 dB port isolation over the wide operating frequency range. This element also needs a box-shaped reflector to reduce the frequency variation of the radiation patterns and to enhance the gain of the antenna.

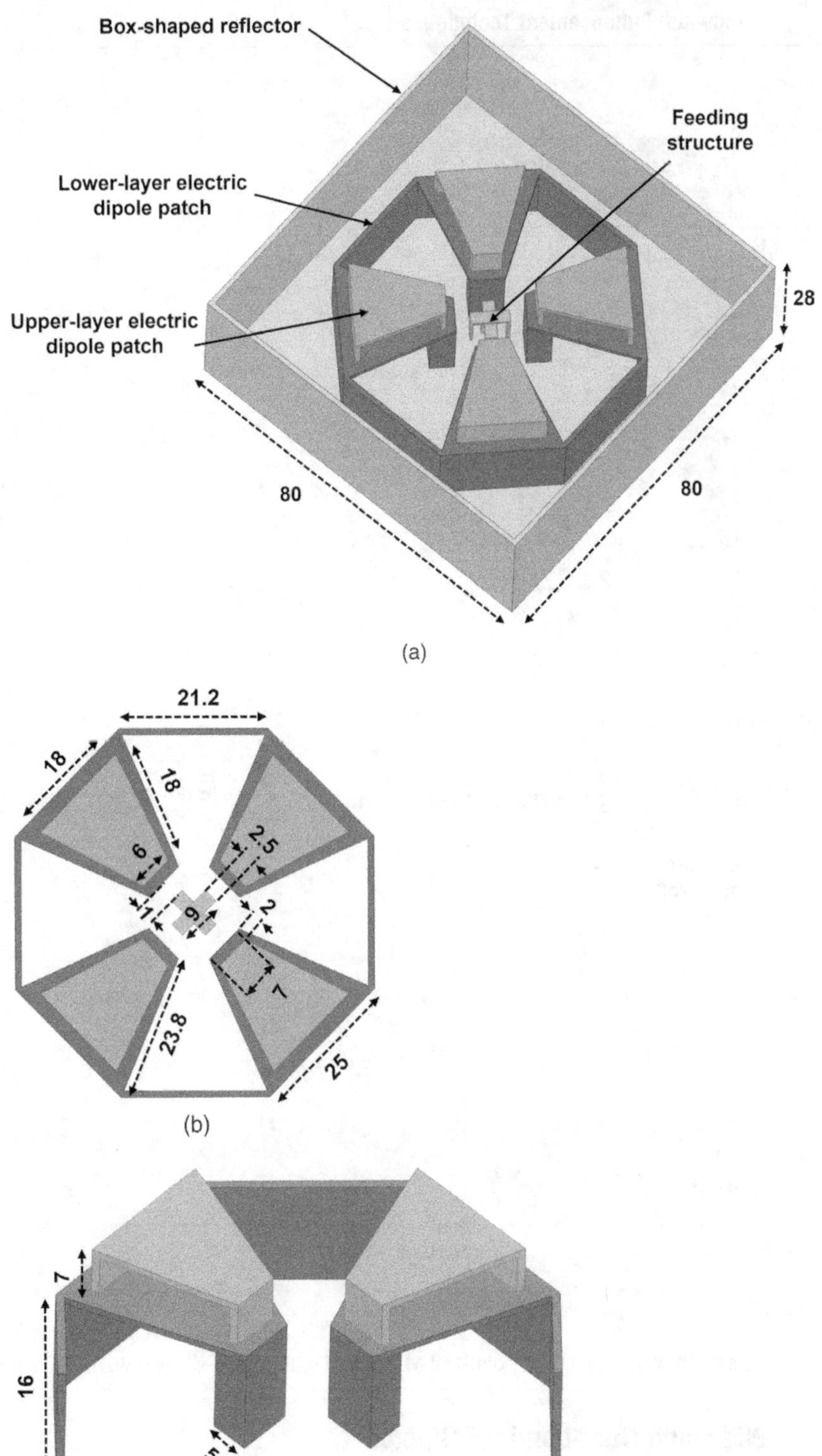

Figure 6.12 Wideband dual-polarized ME dipole with stacked electric dipoles – version 2. (a) Perspective view, (b) Top view, (c) Closer perspective view, and (d) Feed structure. Adapted from [8]

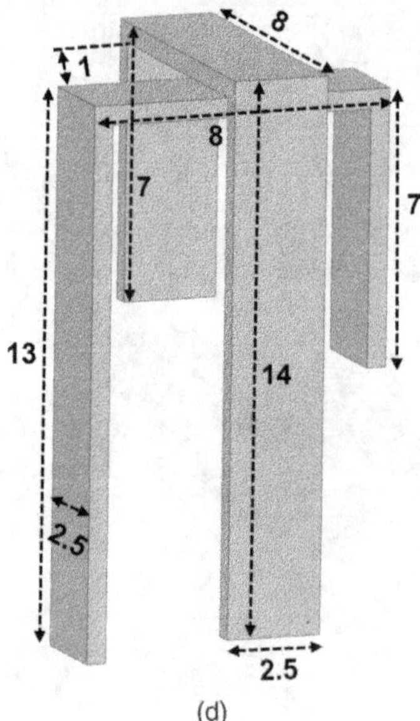

(d)

Figure 6.12 (cont.)

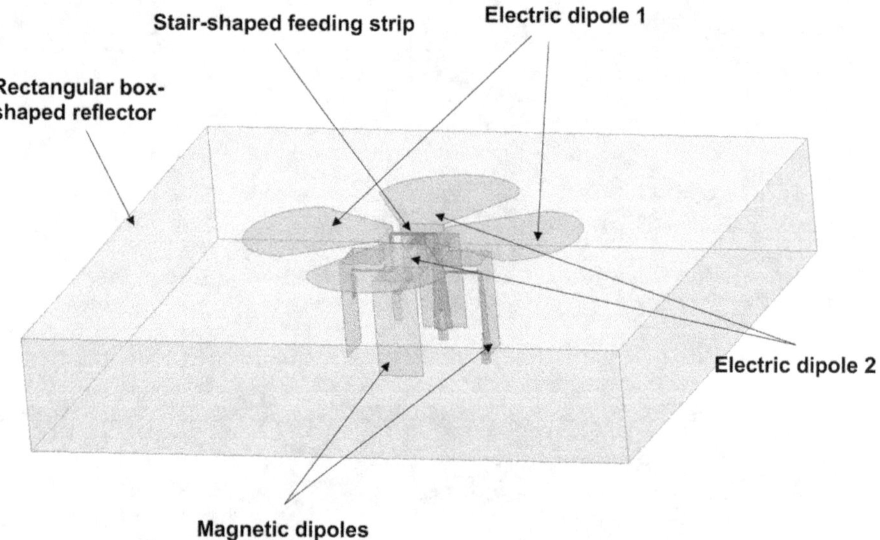

Figure 6.13 Wideband dual-polarized ME dipole with folded vertical walls. Adapted from [9]

6.4   Wideband Dual-band ME Dipoles

Dual-wideband antennas are demanded for base stations covering multiple frequency bands, including PCS, WLAN, WiMAX, 2G, 3G, and LTE. A simple way of generating two wide operating bands for the ME dipole is shown in Figure 6.14. In this design [10], a large slot is introduced to each arm of the planar electric dipole for exciting

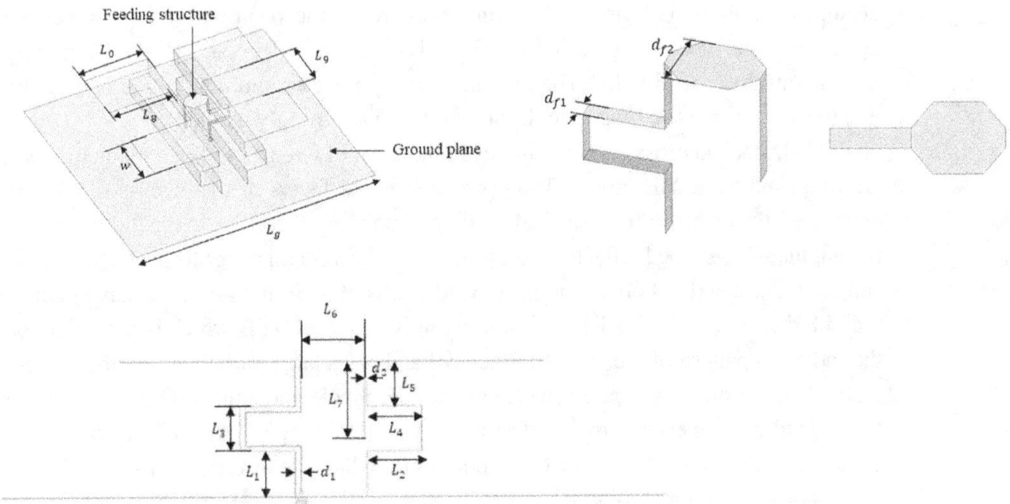

Figure 6.14 A simple dual-wideband ME dipole. Adapted from [10]

an additional radiating mode at higher frequency and the vertical walls are folded for reducing its height. By modifying the shape of the feedline, a dual wideband structure can be achieved amazingly. The prototype with dimensions shown in the figure exhibits 72% bandwidth in the lower band with SWR < 2 from 1.48 to 3.15 GHz and 21% impedance bandwidth in the upper band with SWR < 2 from 4.67 to 5.78 GHz. The antenna has low cross polarization and low back radiation over the operating frequencies, demonstrating its attractiveness as an excellent unidirectional antenna. The limitation in this design is its large variation in gain, over 2.5 dB in the upper band. Also, the location of the upper frequency band is not easily adjusted or controlled.

Another approach to realize double frequency bands with wide bandwidth is to employ stacked electric dipoles. A number of interesting designs with linearly polarized radiation were reported [11, 12, 13], which demonstrate the flexibility in adjusting the bandwidth of location of each band using this method. A disadvantage of these design is the radiation pattern in the upper frequency band which can be improved by using a horn-shaped reflector together with a parasitic director [13]. Moreover, the stacked electric dipole can be further enhanced to become a multiband wideband unidirectional antenna with stable radiation pattern by folding the upper and lower electric dipoles appropriately [14].

6.5   Wideband Dual-band Dual-polarized ME Dipoles

Wideband dual-band dual-polarized antennas are more attractive and useful in mobile communications. A number of designs based on ME dipoles with good port isolation ($S_{12} < -25$ dB) were proposed in the literature for possible applications as base station antennas for 3G/4G/LTE/5G/WiMAX/WLAN communication systems.

In [15], stacked cross electric dipoles in the form of U-shaped strips are used for a ME dipole to achieve two wide bands with wide beamwidth. A slant tetrahedral grounded reflector is employed to further increase the beamwidth. The lower and upper frequency bands are centered at 3.5 GHz and 4.9 GHz, respectively. Both bands have about 20% bandwidth. The antenna has high back radiation, which is probably due to the fact that the magnetic dipole mode is not strongly excited.

In [16], stacked cross electric dipoles in the form of two pairs of rectangular patches are proposed for a ME dipole. The upper and lower layers of the electric dipoles are connected through the four vertical walls for realizing the cross magnetic dipoles. A rectangular box-shaped reflector is employed. The lower and upper frequency bands are centered at 2.2 and 5.3 GHz, and have bandwidths of 47% and 30%, and have gains of 9 and 8 dBi, respectively. The back radiation is below −20 dB which is attractive, but the radiation pattern of the upper frequency band has a large variation over the angles.

In [17], the dual-band performance of the ME dipole is obtained through the use of fan-shaped cross electric dipoles. The novelty in this design is that a U-shaped slot is etched in each arm of the cross electric dipoles to reduce the lower band frequencies and a parasitic circular patch is added above the center of the cross electric dipoles to reduce the upper band frequencies. This would effectively decrease the footprint of the antenna. The antenna has 52% bandwidth at the lower frequency band from 2.14 to 3.66 GHz, and only 7.4% bandwidth at the upper frequency band from 4.68 to 5.04 GHz. Similar to the previous one, a box-shaped reflector is needed to suppress the backside radiation, but the radiation pattern may not be acceptable as it has a large variation over the angles.

In [18], it is demonstrated that if the vertical walls of the ME dipole are folded to become stair-shaped structures together with folded feedlines, wideband dual-band performance can be achieved, with the lower band operating from 1.54 to 2.87 GHz (60% in bandwidth) and the upper band operating from 4.62 to 6.10 GHz (27% in bandwidth). As other designs, a box-shaped reflector is used to reduce the backside radiation. The radiation pattern of this design is much better than the previous two designs, particularly at the upper frequency band.

6.6   Ultra-wideband ME Dipoles

A number of ME dipoles with over 100% bandwidth have been developed for UWB applications. The first one is a modification of the design by Ge and Luk [6] as depicted in Figure 6.8. By introducing a broadband microstrip-to-stripline transition to the feed [19], the antenna can be improved in bandwidth to about 110% with SWR < 2 from 3.08 to 10.6 GHz. A stable gain of 8.7 dBi with variation of 1.9 dB was achieved. In this design, as shown in Figure 6.15, four connected vertical side walls were added to the rectangular grounded plane to improve the backside radiation and maintain a stable gain over the ultrawide bandwidth, as depicted in Figure 6.16.

Several attempts to further improve the performance of the UWB ME dipoles can be found in the literature. In [20] by Botao Feng et al., a modified horn reflector and a U-shaped bowtie electric dipole were used to reduce the size of the UWB ME dipole. A design having an impedance bandwidth of 118% with SWR < 2 from 2.81 to 10.92 GHz and a stable gain of 10 dBi with ±1.2 dB variation was reported. In [21] by Shu-Xi Gong's group, a vertical rhombic loop was used to combine with a horizontal

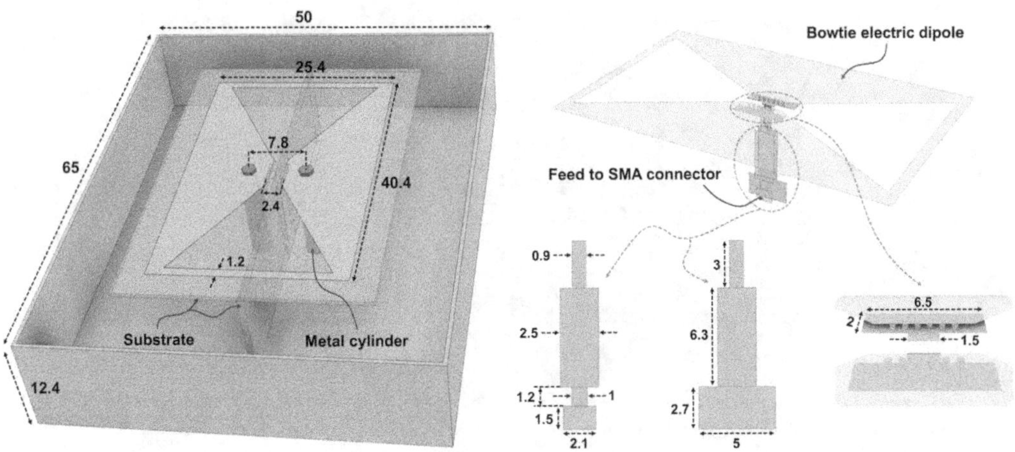

Figure 6.15 UWB ME dipole with improved feed. Adapted from [19]

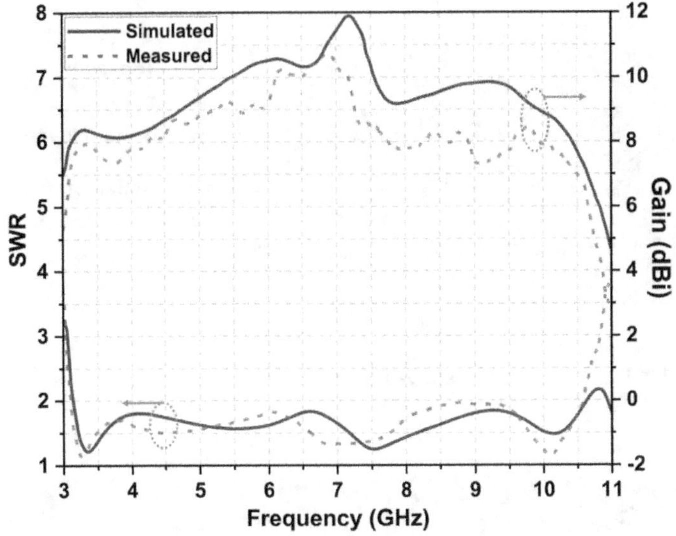

Figure 6.16 Performance of UWB ME dipole with improved feed. Adapted from [19]

bowtie electric dipole to achieve a UWB ME dipole exhibiting an impedance bandwidth of 121% with SWR < 2 from 3.5 to 14.35 GHz. The gain varies between 7.7 and 9.8 dBi over the operating frequencies.

The above UWB antennas have a critical issue. Their radiation patterns are not stable over the broad frequency range. In particular, they have high sidelobe levels at higher operating frequencies. This is probably due to the excitation of higher-order modes in the antenna. To solve this issue, Chang-Hong Liang's group [22] proposed to use vertical shorting pins at the four corners of the ground plane to stabilize the variation of radiation pattern over the operating frequencies, as shown in Figure 6.17. This modified grounding structure is later called a claw-shaped reflector [23]. The antenna, as shown in Figure 6.18, is simple in fabrication and compact in size as it is

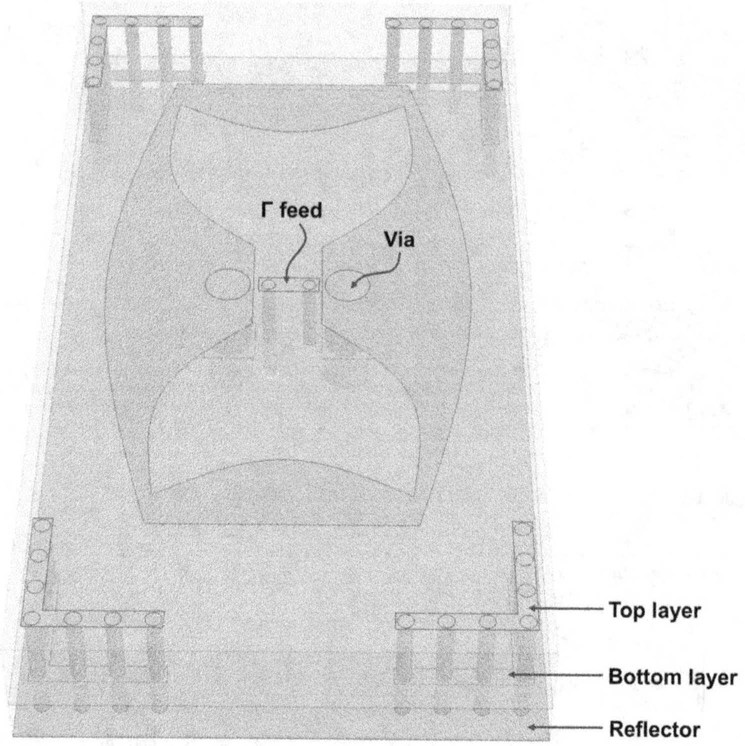

Figure 6.17 Substrate-integrated UWB ME dipole. Adapted from [22]

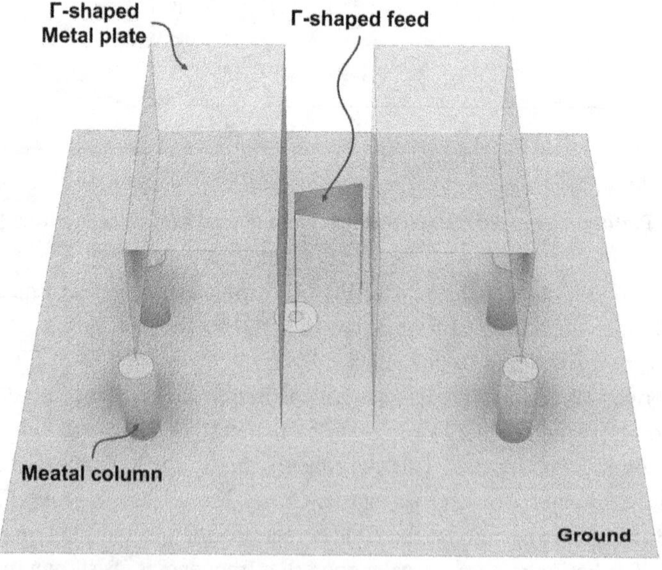

Figure 6.18 ME dipole with wide bandwidth. Adapted from [28]

integrated to dielectric substrates, making it suitable for point-to-point communications, security transmission, and ground-penetrating radar applications.

For further reducing the cross-polarization levels and achieving very symmetrical radiation patterns, several differentially fed ME dipoles have been proposed [24, 25, 26]. In general, the feeding structure is simpler and easier to fabricate for this kind of design. A compact differentially fed UWB ME dipole for medical telemetry can be found in the literature [27].

6.7   Wideband ME Dipoles with Wide Beamwidth

Wideband antennas with wide beamwidth are required for cellular communications with wide-angle coverage, radars with target tracking angles, and others. They are also in great demand for developing beam-steerable antenna arrays for millimeter-wave and terahertz applications. A number of techniques for increasing the beamwidth of ME dipoles have been proposed in the literature. In [28], a folded electric dipole with the edges shorted to the ground through several shorting pins are employed. With appropriate dimensions, a prototype with 81% impedance bandwidth, SWR < 2 from 3.3 to 7.8 GHz, was presented. The antenna has a half-power beamwidth of 215 and 186 in the E-plane and H-plane, respectively at 5.5 GHz (Figure 6.19).

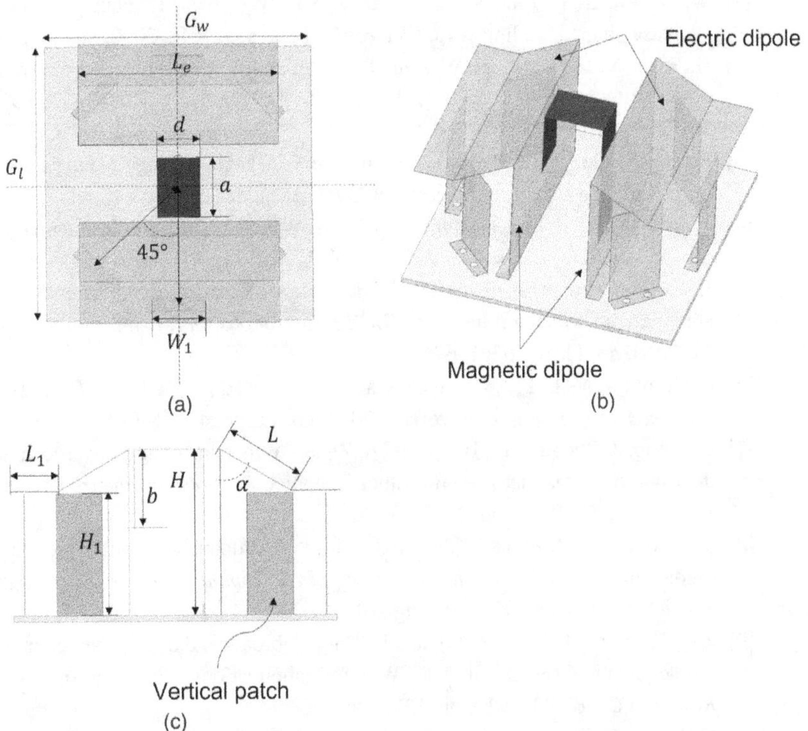

Figure 6.19 ME dipole with wide bandwidth. (a) Top view, (b) Perspective view, and (c) Side view. Adapted from [29]

Another innovative version was proposed by Prof. X. Y. Cao's group, in which the edges of the electric dipole are shorted virtually to the ground through six parasitic vertical strips. In this design, the electric dipole is tilted and bent to help achieve wide beamwidth. The antenna has 55% impedance bandwidth, with SWR < 2 from 2.3 to 4.2 GHz. The beamwidth is about 120 degree in both E-plane and H-plane. Due to its small size, it was used as a basic element to build a phased array with wide scan angle of ±45 degree with less than 3 dB loss [29, 30].

6.8   Summary

Most of the available and effective designs for ME dipoles with wide bandwidth or ultrawide bandwidth are reviewed. Although mainly operating at lower microwave frequencies, these designs can be modified and enhanced for millimeter wave applications with advanced fabrication processes, including low-cost 3D printing technologies. More details can be found in Chapter 7.

References

[1] W. X. An, K. L. Lau, S. F. Li and Q. Xue, Wideband E-shaped dipole antenna with staircase-shaped feeding strip, *Electronics Letters*, vol. 46, 2010, no. 24, pp. 1583–1584.

[2] H. Yuan, X. Li, X. Shuai, W. Ren, T. Zhang and D. Shen, A high-gain magneto-electric dipole antenna, *Proceedings of Asia-Pacific Conference on Antennas and Propagation*, 2017.

[3] K. He, S. Gong and F. Gao, Low-profile wideband unidirectional patch antenna with improved feed structure, *Electronics Letters*, vol. 51, 2015, no. 4, pp. 317–319.

[4] J. Zeng and K. M. Luk, A simple wideband magnetoelectric dipole antenna with a defected ground structure, *Electronics Letters*, vol. 17, 2018, no. 8, pp. 1497–1499.

[5] J. Tao, Q. Feng and G. A. E. Vandenbosch, Director-loaded magneto-electric dipole antenna with wideband flat gain, *IEEE Transactions on Antennas and Propagation*, vol. 67, 2019, no. 11, pp. 6761–6769.

[6] L. Ge and K. M. Luk, A wideband magneto-electric dipole antenna, *IEEE Transactions on Antennas and Propagation*, vol. 60, 2012, no. 11, pp. 4987–4991.

[7] C. Wang, Z. Zheng, J. Li, H. Shi and A. Zhang, Wideband double-layered dielectric-loaded dual-polarized magneto-electric dipole antenna, *Progress in Electromagnetics Research Letters*, vol. 63, 2016, pp. 23–28.

[8] Y. Guo, C. Gu, J. Zhu, X. Chen and Z. Li, A broadband dual-polarized magneto-electric dipole antenna, *Proceedings of the Applied Computational Electromagnetic Society Symposium*, Suzhou, 99–100, Aug 2017.

[9] Z. Li, Y. Sun, M. Yang, Z. Wu and P. Tang, A broadband dual-polarized magneto-electric dipole antenna for 2G/3G/LTE/WiMAX applications, *Progress in Electromagnetics Research C*, vol. 73, 2017, pp. 127–136.

[10] K. He, S. X. Gong and F. Gao, A wideband dual-band magneto-electric dipole antenna with improved feeding structure, *IEEE Antennas and Wireless Propagation Letters*, vol. 13, 2014, pp. 1729–1732.

[11] B. Feng, W. Hong, S. Li, W. An and S. Yin, A dual-wideband double-layer magneto-electric dipole antenna with a modified horned reflector for 2G/3G/LTE applications, *International Journal of Antennas and Propagation*, vol. 2013, article ID 509589.

[12] B. Feng, W. An, S. Yin, L. Deng and S. Li, Dual-wideband complementary antenna with a dual-layer cross-ME-dipole structure for 2G/3G/LTE/LAN applications, *IEEE Antennas and Wireless Propagation Letters*, vol. 14, 2015, pp. 626–629.

[13] J. Tao, Q. Feng and T. Liu, Dual-wideband magnetoelectric dipole antenna with director loaded, *IEEE Antennas and Propagation Letters*, vol. 17, 2018, no. 10, pp. 1885–1889.

[14] G. Yang, S. Zhang, J. Li, Y. Zhang and G.F. Pedersen, A multi-band magneto-electric dipole antenna with wide beam-width, *IEEE Access*, vol. 8, 2020, p. 68821.

[15] B. Feng, C. Zhu, J. C. Cheng, C. Y. D. Sim and X. Wen, A dual-wideband dual-polarized magneto-electric dipole antenna with dual wide beamwidths for 5G MIMO microcell applications, *IEEE Access*, vol. 7, 2019, p. 43346.

[16] M. Yang, J. Zhou, W. Liau and B. Chen, Dual-band dual-polarized magneto-electric dipole antenna with dual-layer structure, *Progress in Electromagnetics Research*, vol. 92, 2020, pp. 193–202.

[17] C. Zhu, B. Wang, W. Luo, H. Hao and P. Wang, Dual-wideband dual-polarised magnetoelectric dipole antenna for 4G/5G microcell base station, *Electronics Letters*, vol. 56, 2020, no. 6, pp. 269–271.

[18] F. Li, Y. Sun, H. Zhu, J. Yu and Y. Fang, A novel wideband dual-band dual-polarized magneto-electric dipole antenna, *Progress in Electromagnetics Research M*, vol. 74, 2018, pp. 73–82.

[19] L. Ge and K. M. Luk, A magneto-electric dipole for unidirectional UWB communications, *IEEE Transactions on Antennas and Propagation*, vol. 61, 2013, no. 11, pp. 5762–5765.

[20] B. Feng, S. Li, W. An, S. Yin, J. Li and T. Qiu, U-shaped bow-tie magneto-electric dipole antenna with a modified horned reflector for ultra-wideband applications, *IET Microwaves, Antennas and Propagation*, vol. 8, 2014, no. 12, pp. 990–998.

[21] K. He, P. Fei and S. X. Gong, A very wideband dipole-loop composite patch antenna with simple feed, *Progress in Electromagnetics Research Letters*, vol. 60, 2016, pp. 9–16.

[22] K. Kang, Y. Shi and C. H. Liang, Substrate integrated magneto-electric dipole for UWB applications, *IEEE Antennas and Wireless Propagation Letters*, vol. 16, 2017, pp. 948–951.

[23] X. Zhang, Y. C. Jiao, Z. B. Weng, Y. X. Zhang and S. Feng, Wideband magneto-electric dipole antenna with a claw shaped reflector for 5G communication systems, *Microwave and Optical Technology Letters*, vol. 61, 2019, pp. 2098–2104.

[24] M. Li and K. M. Luk, A differential-fed magneto-electric dipole antenna for UWB applications, *IEEE Transactions on Antennas and Propagation*, vol. 61, 2013, no. 1, pp. 92–98.

[25] N. Marwah, G. P. Pandey, V. N. Tiwari and S. S. Marwah, A wideband magneto-electric dipole antenna with improved feeding structure, *Advanced Electromagnetics*, vol. 5, 2016, no. 2, pp. 10–16.

[26] C. Y. Shuai and G. M. Wang, A simple ultra-wideband magneto-electric dipole antenna with high gain, *Frequenz*, vol. 72, 2018, pp. 27–32.

[27] J. I. E. Anosike, L. Y. Feng, H. X. Zheng, Y. Liu and Y. X. Liu, Unidirectional UWB magneto-electric antenna for medical telemetry, *Progress in Electromagnetics Research M*, vol. 64, 2018, pp. 211–217.

[28] G. Yang, J. Li, J. Yang and S. G. Zhou, A wide beamwidth and wideband magnetoelectric dipole antenna, *IEEE Transactions on Antennas and Propagation*, vol. 66, 2018, no. 12, pp. 6724–6732.

[29] C. -H. Zhang, X. -Y. Cao, J. Gao and S. -Jia. Li, Broadband and wide beam magneto-electric dipole antenna design, *Journal of Electronic & Information Technology*, vol. 37, 2015, no. 3, pp. 758–762.

[30] C. Zhang, X. Y. Cao, J. Gao and J. F. Han, Broadband and wide beam magneto-electric antenna and its application in scanning array, *2018 International Conference on Microwave and Millimeter Wave Technology*, 2018.

7 Millimeter-wave Magnetoelectric Dipoles

7.1 Introduction

In the era of the fifth Generation (5G) of wireless communications, many interesting applications operating at the millimeter-wave band, including 28, 37, 39, and 64–71 GHz, are being developed, attributing to the advantages of wider spectral resource and higher data throughput compared with current communication systems working at or below 6 GHz. To tackle the high propagation loss at millimeter waves, the multi-input multi-output (MIMO) arrays or beam steerable arrays have been proposed. The MIMO arrays can form multiple beams pointing to multiple users at different directions, which is an enhancement of the beam steerable arrays. Efficient wideband antenna arrays are in high demand to support this brilliant growth in wireless technologies.

The magnetoelectric (ME) dipole was initially developed for use at lower microwave frequencies. It was exciting to find that the antenna can be integrated into printed circuit board (PCB) for millimeter-wave (mmWave) applications. Due to its high antenna efficiency and its suitability for antenna array realization, much effort has been spent by many antenna research and development groups across the globe to enhance the performance of mmWave ME dipoles, facing the need to achieve numerous wireless communication systems operating at the millimeter-wave and terahertz frequency bands.

In this chapter, a comprehensive review on using different transmission lines for feeding ME dipole antennas and arrays is presented, including the substrate integrated waveguide (SIW), ridge gap waveguide, packaged microstrip line, and substrate integrated coaxial line (SICL) feeds. In addition, the developments of low profile of ME dipole arrays, filtering ME dipoles, and all-metal ME dipole arrays for high-power applications are summarized. Some other recent applications are briefly reported. Hopefully, our readers can appreciate the attractiveness of the ME dipoles for future wireless applications at millimeter-wave and terahertz frequencies.

7.2 Basic Designs

The first millimeter-wave ME dipole proposed by K. B. Ng et al. [1] is shown in Figure 7.1. The planar electric dipole is printed on the top surface of a grounded dielectric substrate with thickness of H = 0.787 mm and ε_r = 2.2. To realize the magnetic dipole, three vertically oriented metallic vias, instead of a vertical wall, are added under the inner edge of each arm of the planar electric dipole. All the six vias have one end in contact with the planar dipole and the other end in contact with the grounded plane, except the center one located under one arm for the excitation of the antenna. This particular via, together with the other two neighboring shorted vias, realizes a transmission line. One end of this via is connected to a microwave launcher and the other end is connected to the T-shaped feed printed on the dielectric substrate. The T-shaped feed is more easily to be realized than the L-probe feed commonly used at lower microwave frequencies. The ends of the T-shaped strips are slightly enlarged in width to avoid possible peeling off from the dielectric substrate.

The thickness of the dielectric substrate is about one quarter of the wavelength in dielectric at 60 GHz. This makes it suitable for the design of an ME dipole at this frequency range. The total length of the electric dipole, $L = 2L_1 + L_2$, was selected at 2.2 mm, which is about 0.5 wavelength in dielectric. The width of the electric dipole was chosen as 1.4 mm to achieve good radiation characteristics.

A prototype was made and tested. From Figure 7.2(a), it can be observed that the antenna exhibits 43% impedance bandwidth with $S_{11} < -10$ dB from 48 to 75 GHz. The average gain is about 7.5 dBi over the frequency range from 50 to 70 GHz, as

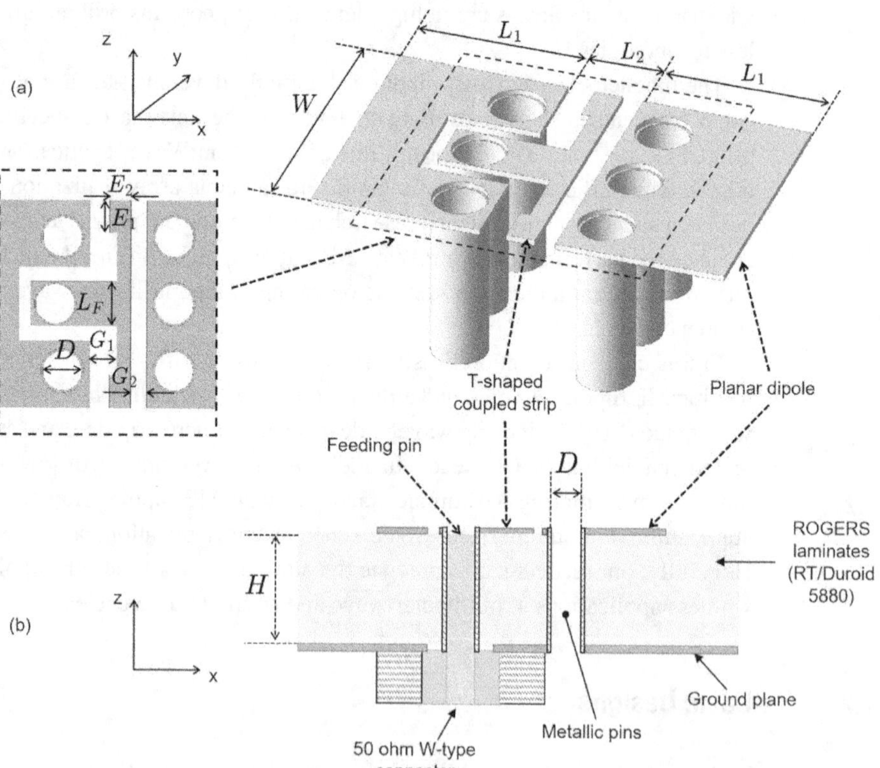

Figure 7.1 The first mmWave ME dipole. (a) 3D view and (b) Side view [1]

shown in Figure 7.2(b). It was also found that although the T-probe feed was used instead of the L-probe feed, the cross polarization can still be lower than −15 dB over the operating frequencies. The performance of this millimeter-wave ME dipole is comparable to those lower frequency designs. It confirms that the loss due to surface wave excitation is not significant. This encouraging finding has motivated many new applications of the ME dipoles at millimeter-wave and terahertz frequencies.

The T-shaped feed can be fabricated at low cost and can help achieve wideband characteristic, but the cross polarization and back radiation are higher than the design with L-probe feed. Since realizing an L-probe feed is complicated in PCBs, a modified design based on a proximity-coupled feed was proposed by M. Li et al. for achieving better performance in 2015 [2].

As depicted in Figure 7.3, the antenna has four rectangular patches printed on the top surface of a grounded dielectric substrate, with each patch supported by three vertical vias of radius $r = 0.15$ mm located at the inner edges of the patch. The thickness of the dielectric substrate is $t = 0.787$ mm, which is equivalent to a quarter of a

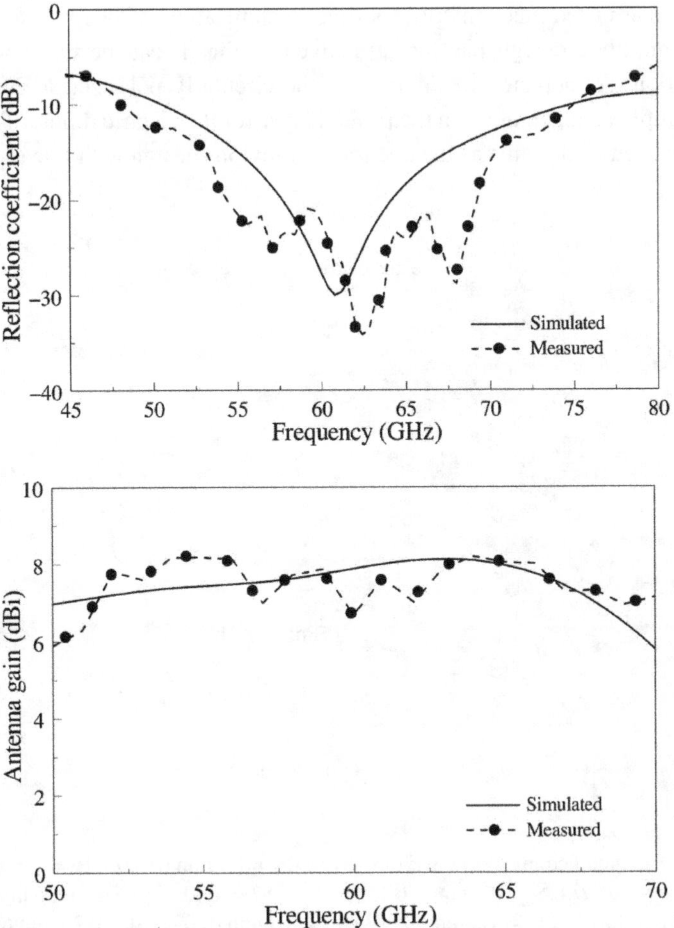

Figure 7.2 Simulated and measured results. (a) Reflection coefficient and (b) Antenna gain [1]

wavelength at around 60 GHz. The dielectric constant of the supporting substrate is $\varepsilon_r = 2.2$. The four planar patches together function as an electric dipole and the vertical vias together with the portion of the grounded plane between the vias perform as a vertically oriented quarter-wave patch antenna to provide the magnetic current radiation.

The feed is in an L-shape, but its operating principle is different from the one used for exciting the conventional microstrip patch antenna. The vertical portion of the L-probe, together with the neighboring shorting vias, performs as a transmission line, whereas the horizontal portion serves as a feed to couple the signal electromagnetically to the electric and magnetic dipoles.

With dimensions selected after a detailed parametric study, as shown in the figure caption of Figure 7.3, a prototype was designed and fabricated for measurements. It can be seen from Figure 7.4 that the SWR is less than 2 between 41.5 and 69.5 GHz, exhibiting 51% impedance bandwidth. The measured gain is about 8 dBi. Symmetrical radiation patterns with low cross polarization (less than −20 dB) were also observed. Although the projection area of the antenna is slightly larger than the previous one. It still can be used in antenna array designs. Its high-efficiency performance has attracted much interest for diverse applications to be reviewed later.

This modified design has the attractiveness that it can be made with circular polarization. As depicted in Figure 7.5, the circularly polarized ME dipole has a more complicated geometry in the printed layer for the electric dipole. The dielectric substrate used is the same as the one for the previous design with linear polarization.

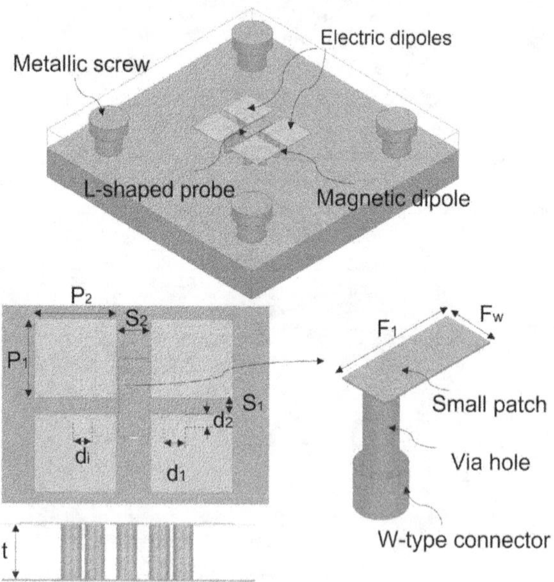

Figure 7.3 The modified mmWave ME dipole [2]. ($P_1 = 1.2$ mm (0.33λ), $P_2 = 1.3$ mm (0.36λ), $S_1 = 0.25$ mm (0.07λ), $S_2 = 0.55$ mm (0.15λ), $t = 0.787$ mm (0.22λ), Fp $= 0.35$ mm (0.1λ), $F_w = 0.49$ mm (0.13λ), $F_1 = 1.2$ mm (0.33λ), $d_1 = 0.3$ mm (0.09λ), $d_i = 0.35$ mm (0.1λ), $d_2 = 0.2$ mm (0.05λ), where λ is wavelength in dielectric at center frequency.)

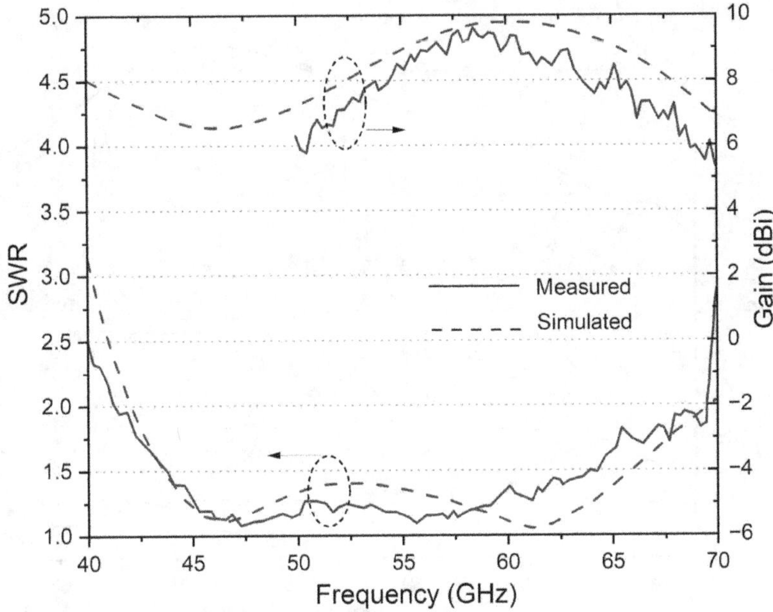

Figure 7.4 Gain and return loss of mmWave ME dipole [2]

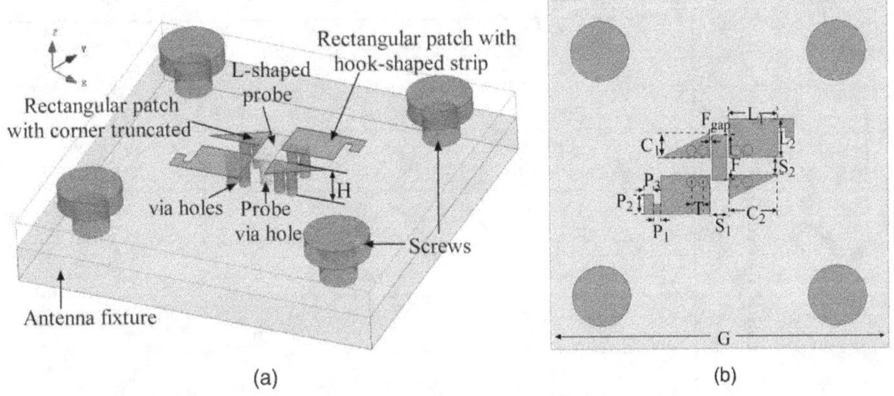

Figure 7.5 Printed circularly polarized ME dipole [3]. (a) Perspective view and (b) Top view. (L_1 = 1.5 ($0.28\lambda_0$), L_2 = 1.05 ($0.19\lambda_0$), S_1 = 0.58 ($0.11\lambda_0$), S_2 = 0.5 ($0.09\lambda_0$), H = 0.787 ($0.15\lambda_0$), P_1 = 0.22 ($0.04\lambda_0$), P_2 = 0.55 ($0.10\lambda_0$), P_3 = 0.27 ($0.05\lambda_0$), C_1 = 0.67 ($0.12\lambda_0$), C_2 = 1.4 ($0.26\lambda_0$), F = 1.3 ($0.24\lambda_0$), F_{gap} = 0.03 ($0.006\lambda_0$), T = 0.35 ($0.06\lambda_0$), D = 0.3 ($0.06\lambda_0$), G = 10 ($1.85\lambda_0$) (unit is in mm))

In this design, two vertical vias are employed to support each patch. The detailed dimensions are shown in the caption of Figure 7.5.

A prototype was constructed for measurements to confirm the simulation results. The measured and simulated impedance bandwidths are 57% (with SWR < 2 from 38.5 to 69 GHz) and 56% (with SWR < 2 from 38.2 to 69.5 GHz), respectively, as shown in Figure 7.6(a). Over the operating frequency range, the simulated gain in

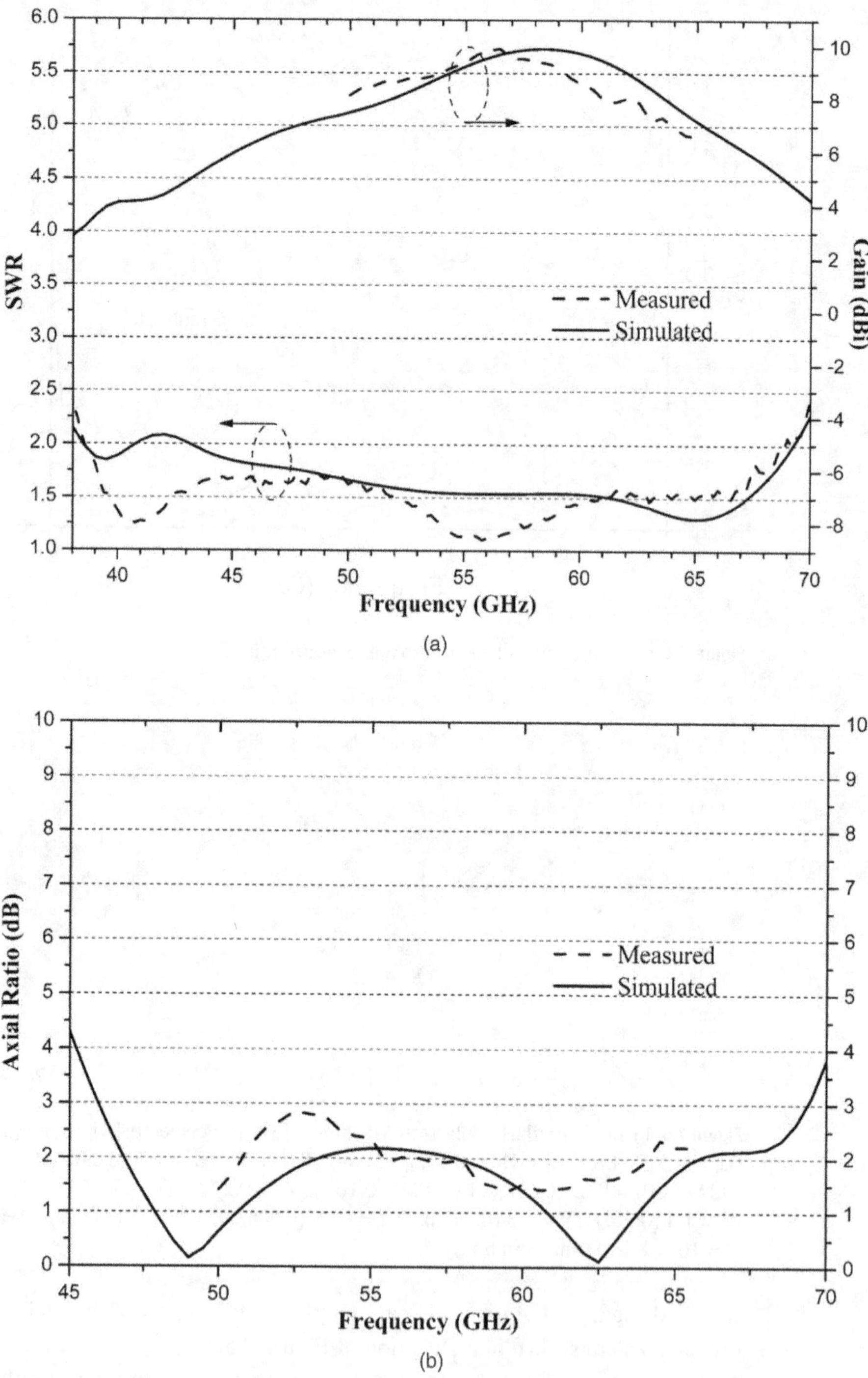

Figure 7.6 Simulated and measured results. (a) Gain and SWR and (b) Axial ratio [3]

the normal direction increases with frequency from 3 dBic at 38.2 GHz to 9.9 dBic at 60 GHz and then decreases to 4 dBic at 79.5 GHz. The measured gain agrees well with the simulation. This behavior is different from the linearly polarized design having almost constant gain over the operating frequencies. As shown in Figure 7.6(b), the simulated axial ratio is less than 3 dB from 45.8 to 69.4 GHz, giving a very wide axial-ratio bandwidth of 41%, which is in good agreement with the measured data. The result demonstrates that a printed mmWave circularly polarized ME dipole with a simple feed can be realized with a very wide axial-ratio bandwidth.

The major advantage of the 4-patch design of the ME dipole is that it can be operated with dual-polarization. Several related works were published in 2019. A dual linearly polarized design is shown in Figure 7.7, which is designed for 5G millimeter-wave MIMO arrays [4]. The design is integrated into a dielectric substrate with $\varepsilon_r = 2.2$. Two crossed L-probes are located in the cross gaps between the vertical vias. With optimized values for the dimensions of the antenna, the achieved impedance bandwidth is 17% with SWR < 2 between 25 and 29.8 GHz, and the isolation is over 25 dB in simulation. The radiation pattern is very symmetrical with low back radiation.

Another similar design was proposed by Prof. Yujian Li and Prof. Yong Xin Guo, as shown in Figure 7.8 [5]. It was designed for fabrication by the LTCC technology. With multiple laminates of ceramic layers, complicated 3-D structures can

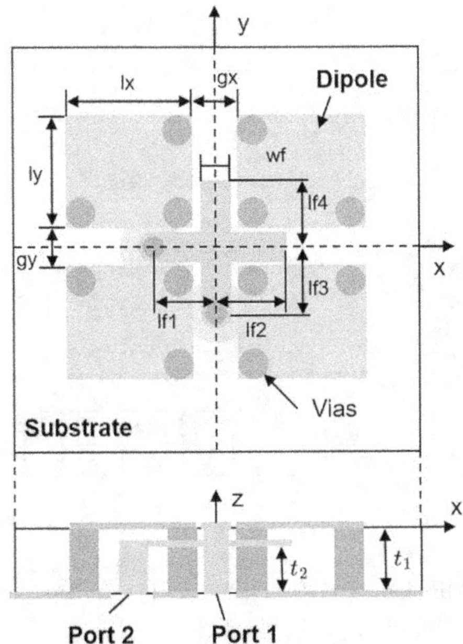

Figure 7.7 A dual-polarized ME dipole for MIMO arrays [4]

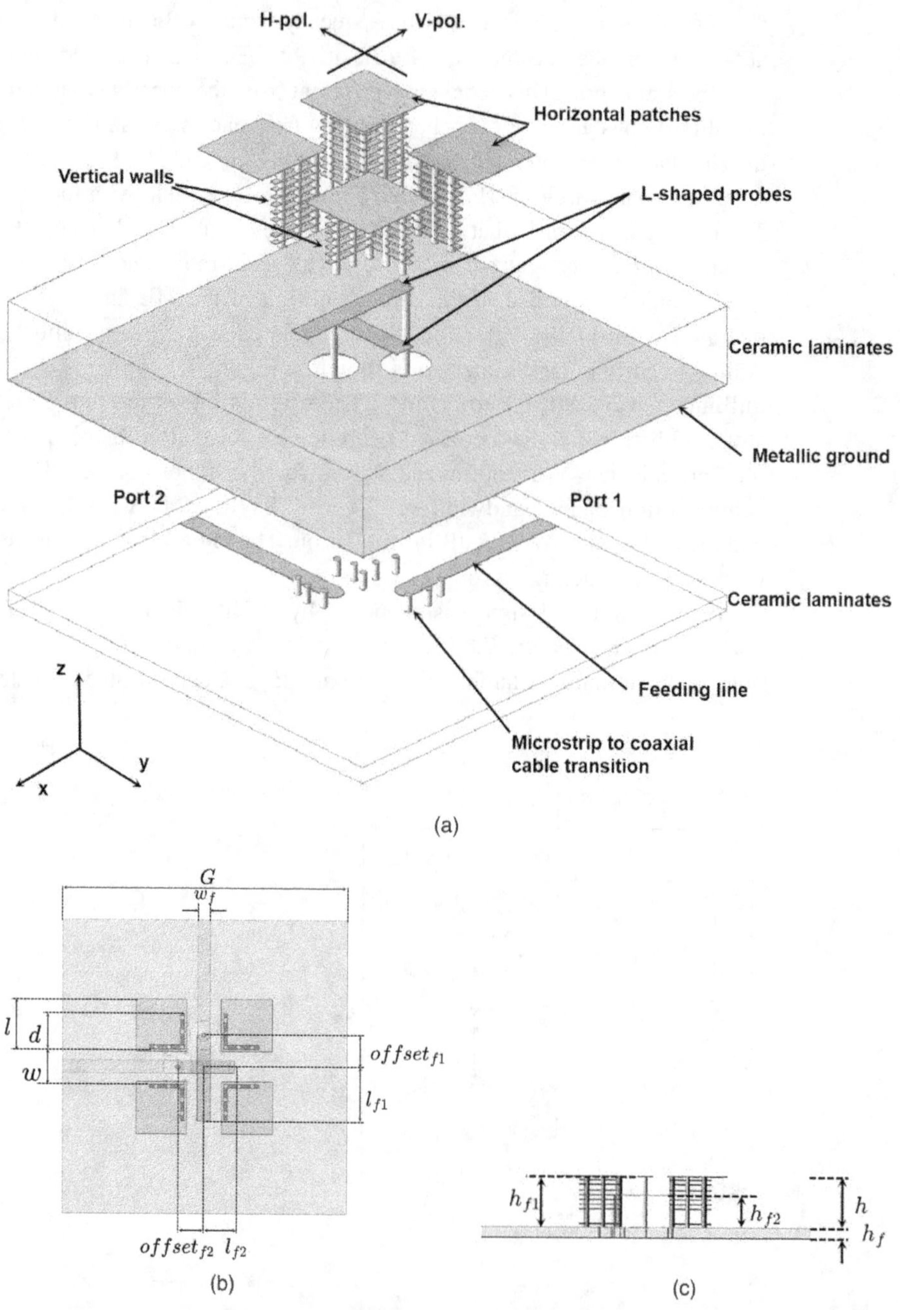

Figure 7.8 Dual linearly polarized ME dipole on LTCC laminates [5]. (a) Perspective view, (b) Top view, and (c) Side view

be realized easily. A design was achieved with 42.5% 3-dB impedance bandwidth, 24 dB isolation between the two input ports, and up to 6.9 dBi gain. A 4 × 4 broadside array based on this antenna element was also designed with a microstrip line feed

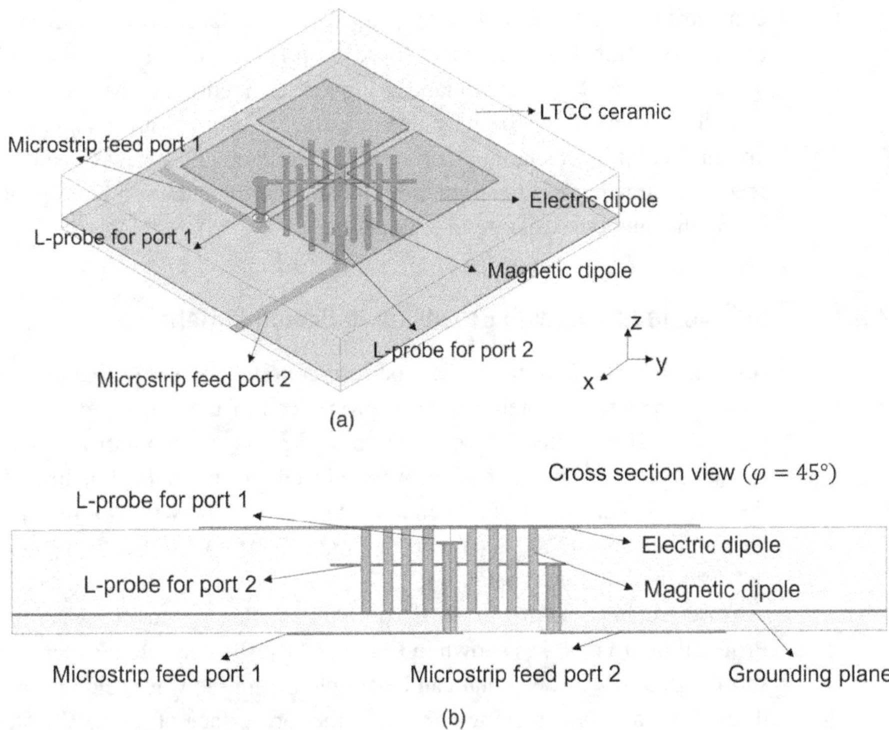

Figure 7.9 Dual-polarized ME dipole with high isolation. (a) Side view (b) Perspective view and [6]

network. Encouraging performance, including 45% impedance bandwidth covering the entire Ka-band, a maximum gain of 16 dBi and 70% radiation efficiency was demonstrated for the prototype made. We notice that the previous design [4] has narrower bandwidth than this one [5]. The reason is probably due to the fact that the vertical vias were not placed at appropriate locations.

The isolation of the above two designs is only around 25 dB. To achieve higher isolation, the feed locations and orientations need to be changed. A possible approach is shown in Figure 7.9, where the vertical portions of the L-probe feeds are located outside the crossed channels formed by the vertical vias [6]. This ME dipole was designed for LTCC fabrication and more than 36 dB isolation over 8% impedance bandwidth with SWR < 2 from 26.4 to 28.6 GHz. The gain of the antenna is about 5.5 dBi. Due to its small size, it is suitable for realizing antenna arrays for 5G millimeter-wave devices.

7.3　SIW-fed ME Dipoles

The class of SIW has been widely applied for designing millimeter-wave circuits and devices owing to its low-loss characteristic and low-cost fabrication by conventional plate-through-hole printed circuit fabrication process. Substantial amount of effort has been spent to develop ME dipole arrays with SIW feed networks. Several high-performance, high-gain fixed-beam ME dipole arrays are described briefly later. In these designs, the ME dipoles are mainly excited by the aperture coupling technique for ease of fabrication, instead of using the L-probe feed. The drawback is the reduction in bandwidth, which can be mitigated by modifying the antenna structure ingeniously.

7.3.1   ME Dipole Arrays with Broadside Radiation Pattern

Inspired by the design of the aperture-coupled microstrip patch antennas and dielectric resonator antennas, the L-probe feed of the ME dipole was replaced by a coupling slot located between the vertical posts. With a narrow rectangular slot, the achieved antenna bandwidth was reduced substantially. For linearly polarized designs, the bandwidth can be enhanced by using a bow-tie slot feed instead of the rectangular slot [7, 8], as shown in Figure 7.10, or by increasing the separation between the horizontal patches [9].

A circularly polarized ME dipole antenna element was proposed by Prof. Yujian Li et al., as shown in Figure 7.11 [10]. The antenna element is fed by a short-ended SIW. The signal can be coupled from the waveguide to the ME dipole through a transverse rectangular slot in the top surface of the SIW instead of using an L-shaped probe feed. The structure of the ME dipole consists of four horizontal patches, each of which is supported by one shorting post or via. The use of one large shorting post instead of three small shorting posts is a trade-off between the performance and fabrication cost. It was discovered that circular polarization can be achieved by adding a narrow metallic strip to connect the inner corners of a pair of diagonal patches. The function of the strip is to help split the resonant frequencies

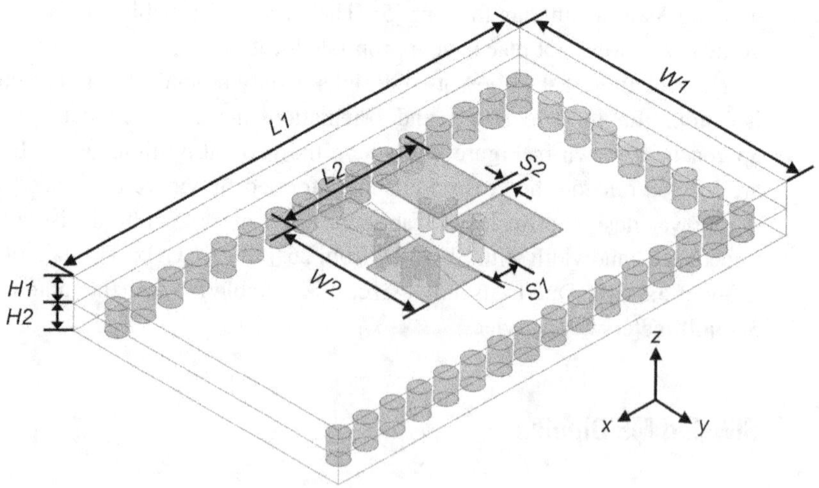

Figure 7.10 Aperture-coupled ME dipole [7]

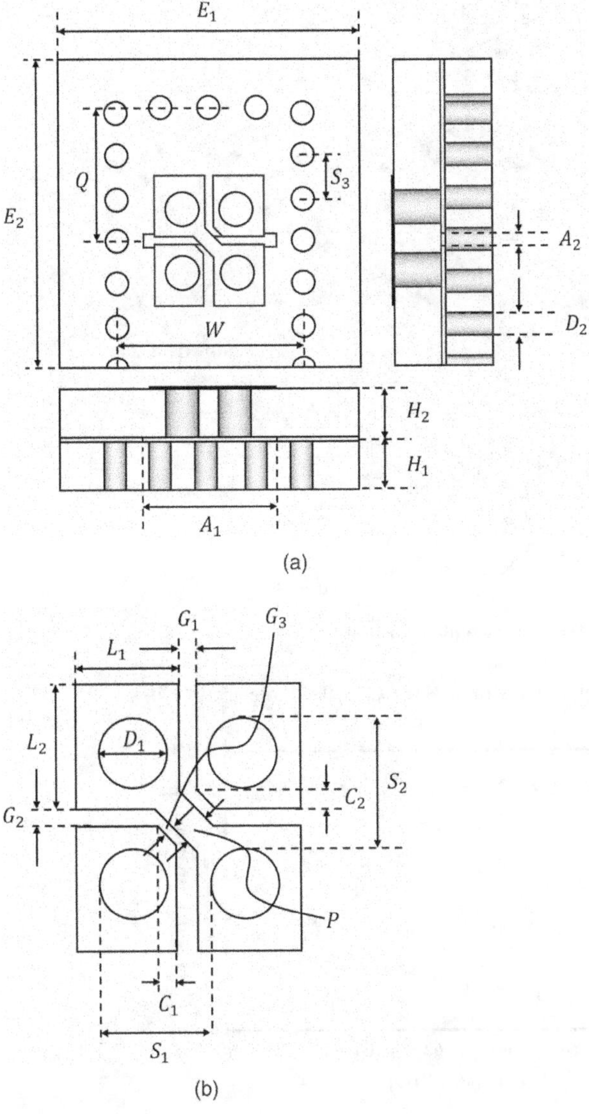

Figure 7.11 Geometry of SIW-aperture-coupled CP ME dipole. (a) Top and side views and (b) Top view of the antenna [10]

of the two orthogonal linearly polarized modes. This design is attractive for antenna array implementation as the size of the antenna element is maintained small, like the linearly polarized design.

With antenna parameters as shown in the figure caption of Figure 7.11, a 60-GHz printed ME dipole element having an impedance bandwidth of 28.8%, an axial-ratio bandwidth of 25.9%, and an average gain of 7.7 dBic was achieved. An 8 × 8 array based on this basic ME dipole structure was designed, as shown in Figure 7.12. The array is fed by a cooperate SIW feed network to achieve low insertion loss.

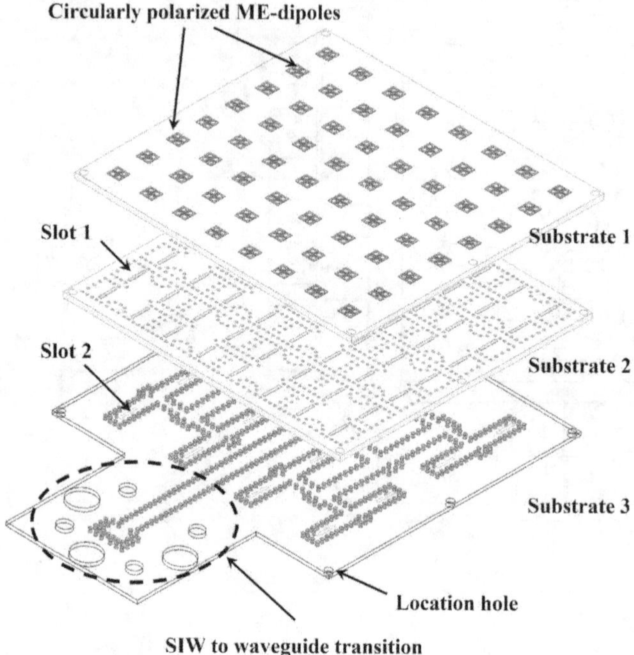

Figure 7.12 Perspective view of an 8 × 8 CP ME dipole array [10]

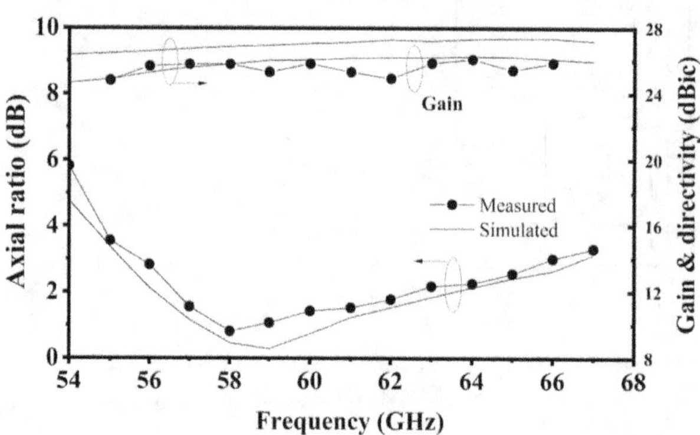

Figure 7.13 Gain, directivity, and axial-ratio versus frequency of CP ME dipole array [10]

The prototype was measured, exhibiting 18.2% impedance bandwidth (SWR < 2), 16.5% axial-ratio bandwidth (AR < 3), 26.1 dBic gain, and 70% radiation efficiency. The antenna gain of the array is very stable over the operating frequencies as depicted in Figure 7.13 [10]. The broadside radiation is acceptable with sidelobes lower than −13 dB and cross polarization lower than −20 dB.

Another aperture-coupled CP ME dipole with SIW feed was also proposed [11]. The antenna is simpler in structure as it is based on the design with two horizontal patches [1]. Circular polarization can be realized by modifying the shape of each horizontal patch. The achieved axial-ratio (AR) bandwidth is limited even using a bow-tie slot feed. If wider AR bandwidth is required, the L-probe feed can be employed to excite the ME dipole with two horizontal patches through the use of an SIW to coaxial transition [12].

As dual-polarized antennas are popularly used nowadays for enhancing communication system performance through the polarization diversity technique, much effort has been spent on the development of dual-polarized ME dipole arrays. The first dual-linearly polarized ME dipole array with SIW feed network is shown in Figure 7.14. It is a 2 × 2 array fed by two stacked SIW beamforming networks. It

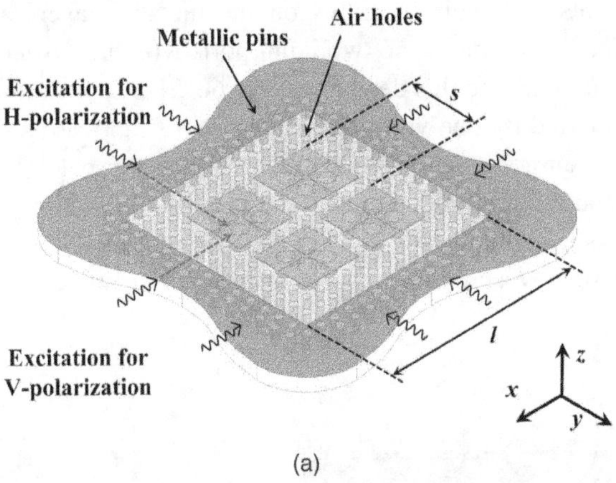

(a)

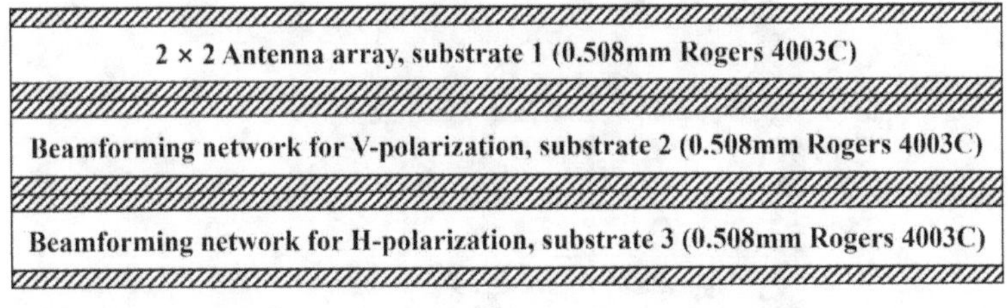

(b)

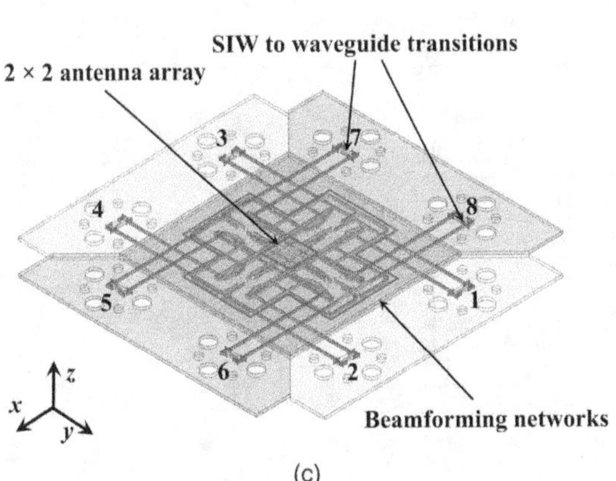

(c)

Figure 7.14 A Multibeam ME dipole array. (a) Perspective view, (b) Side view, and (c) Feed network [13]

is made of three PCBs, with the top layer supporting the ME dipole, the middle layer supporting the feed network for the vertical polarization, and the bottom layer for the horizontal polarization. Each feed network consists of four wideband 90-degree directional couplers specially arranged to excite the four ME dipole elements. For a 2 × 2 array, the four couplers can be printed on a single PCB layer. It would be challenging to design arrays of more elements. One possible approach was reported in [14].

The detailed design of each ME dipole element of the array is shown in Figure 7.15. The slots and their locations on the metallic layers are specially designed to allow the excitation of the two input ports with high isolation of more than 45 dB in simulation. Since this design was for 60 GHz applications, each horizontal patch is supported by one vertical via for ease of fabrication. This change in comparison with the traditional design using three vias per patch, together with the use of aperture coupling technique, makes the antenna difficult to be matched. It was found that the issue could be solved when the inner corners of the four patches were connected together by a crossed metallic strip, as demonstrated in Figure 7.16.

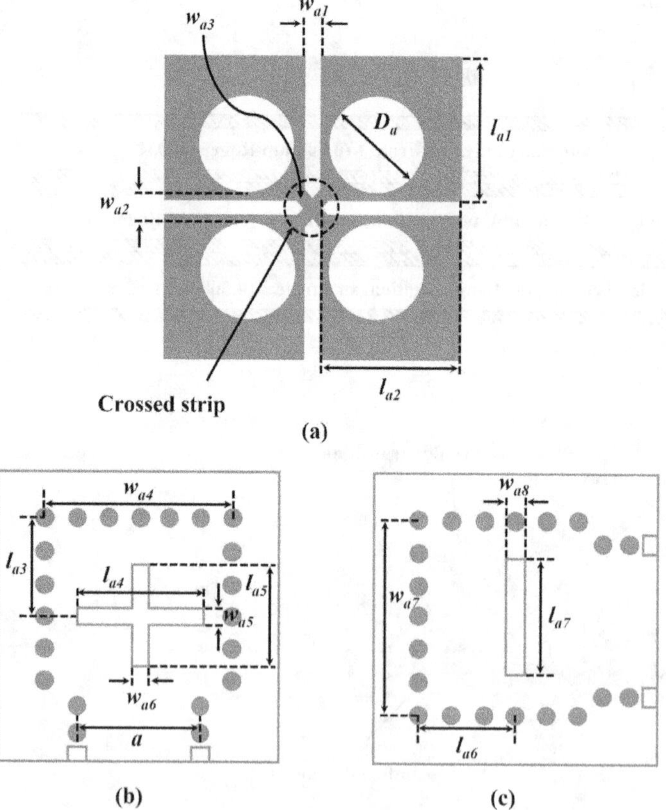

Figure 7.15 Element design of aperture-coupled dual-polarized ME dipole. (a) Top layer, (b) Middle layer, and (c) Bottom layer [13]

With appropriate geometric dimensions and material parameters selected for the design, a prototype was fabricated and tested with 22% impedance bandwidth, 12.5 dBi gain, and 70% antenna efficiency. The element spacing is about 0.6 wavelength in dielectric. The array can radiate four tilted beams as shown in Figure 7.17. The directions in the horizontal planes are 45°, 135°, 225°, and 315°, while the directions of the main beams in the two principal vertical planes are all around 20°.

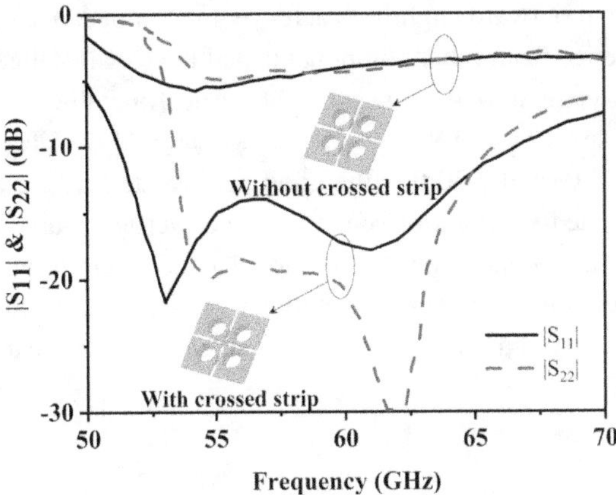

Figure 7.16 Simulated performance of an SIW-fed dual-polarized aperture-coupled ME dipole with and without a crossed metallic strip [13]

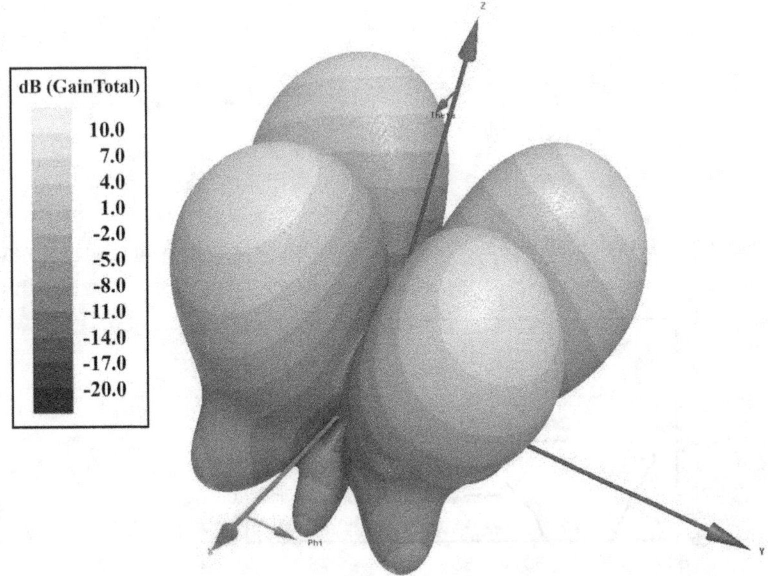

Figure 7.17 Simulated far-field radiation of same polarization of the dual-polarized ME dipole arrays [13]

7.3.2   SIW-fed ME Dipole Arrays with End-Fire Radiation

In some portable devices such as cellular phones and tablets, antenna arrays with multiple end-fire beams of radiation are highly desirable as they can be mounted at the edges of the devices to achieve a wider angle of coverage. The first ME dipole with end-fire radiation is depicted in Figure 7.18. The open-ended SIW radiates as a horizontal magnetic dipole and the two pairs of vertical vias located at the edges of the SIW operate as two closely packed vertical electric dipole. The dielectric substrate is required to extend outward slightly to achieve good impedance matching. To avoid unpredicted influence from other components and devices installed behind the antenna, two U-shaped vertical walls consisting of vertical metallic vias are added around the electric dipoles. For a prototype operated at around 60 GHz [15], it can be observed in Figures 7.19 and 7.20 that the good performance of the end-fire ME dipole is not affected by the incorporation of the U-shaped vertical walls. The antenna element has about 44% impedance bandwidth and 110° beamwidth, which made it highly suitable to realize multi-beam antennas with a wide scan angle over a wide frequency range. Antenna arrays based on this end-fire antenna element can be realized

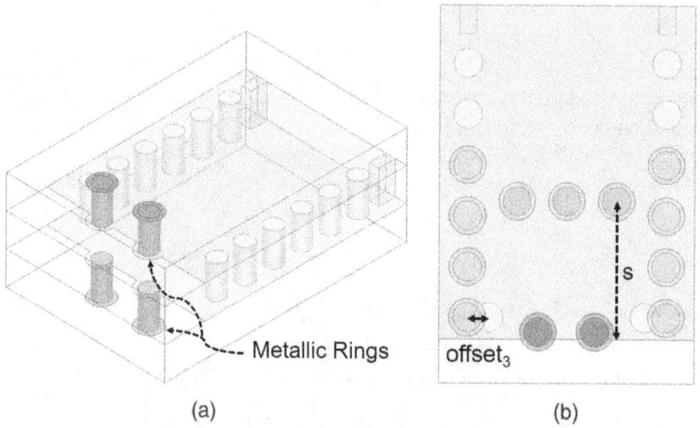

Figure 7.18 End-fire ME dipole fed by SIW. (a) Without and (b) With U-shaped vertical walls [15]

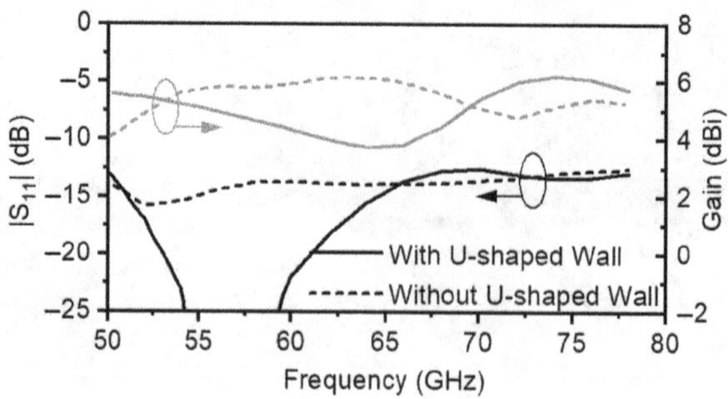

Figure 7.19 Simulated return loss and gain of end-fire ME dipole [15]

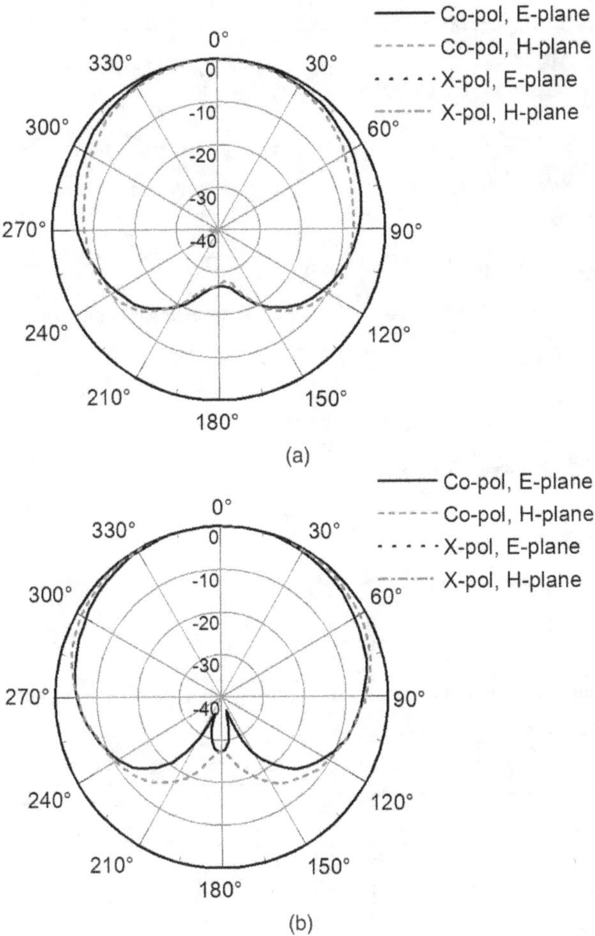

Figure 7.20 Simulated radiation pattern of end-fire ME dipole. (a) Without and (b) With U-shaped vertical walls [15]

at low cost since the array of elements can be fabricated together with the complex feed network in the same PCB.

The above design can only provide vertical polarization radiation. For achieving polarization diversity, end-fire ME dipole antennas and arrays with horizontal polarization were also designed successfully. As shown in Figure 7.21. The basic waveguide structure is integrated into three stacked PCBs. The broad wall of the SIW is realized by vertical vias in the three dielectric layers, while the side walls of the SIW are realized by the metal layers on the upper surface of the top dielectric layer and the lower surface of the bottom layer. The dominant waveguide mode with the electric field in the horizontal direction can be excited. The open end of the waveguide radiates as a vertical magnetic current and the four horizontal rectangular patches added at the open end of the waveguide work as two horizontal electric dipoles. To improve the radiation pattern performance, vertical vias are added behind the electric dipoles

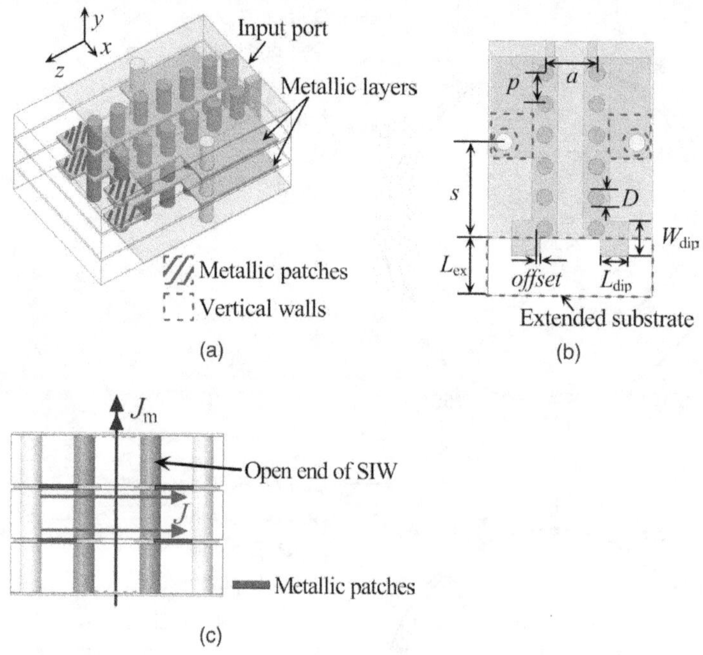

Figure 7. 21 Horizontally polarized end-fire ME dipole. (a) Perspective view, (b) Top view, and (c) Front view [16]

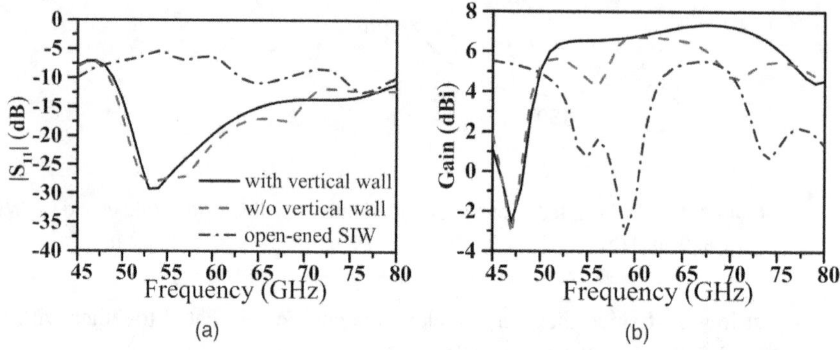

Figure 7.22 Simulated performance of horizontal polarized ME dipole. (a) Return loss and (b) Gain [16]

to form vertical reflecting walls. Similar to the vertically polarized design, the dielectric substrates of this design are also extended outward for improving the matching performance of the antenna. Moreover, an SIW 90-degree twist can be designed to excite this wideband antenna element [16].

A prototype operating at around 60 GHz was designed and simulated. To demonstrate the superiority of the design, the performance of an open-ended SIW, the ME dipole without the vertical wall, and the ME dipole with the vertical wall were studied. As shown in Figure 7.22, the open-ended SIW is poor in impedance matching

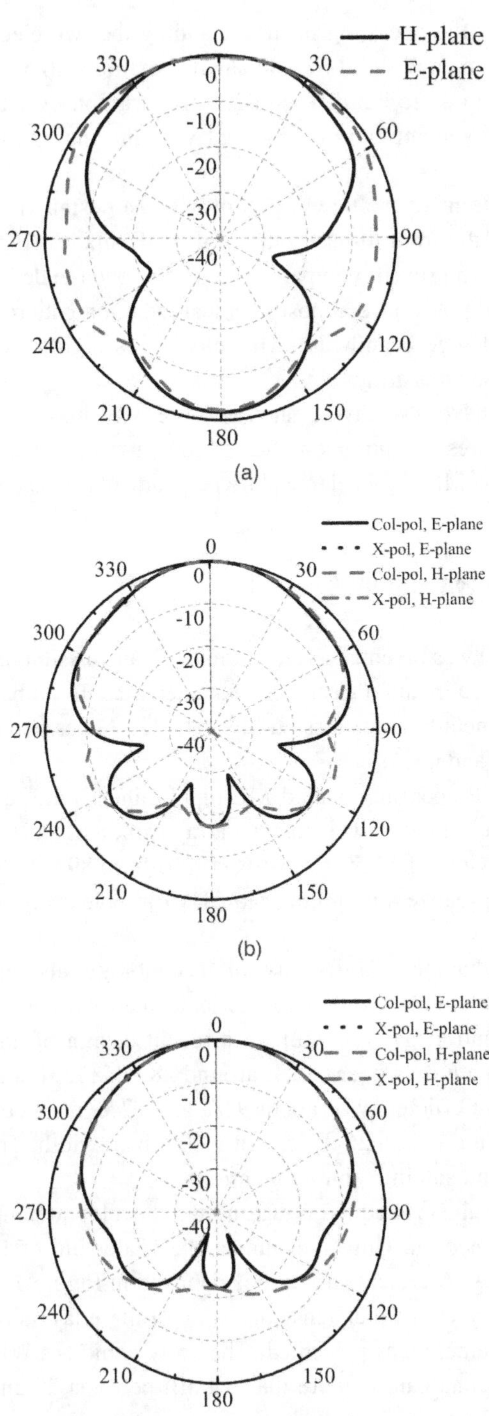

Figure 7.23 Radiation patterns of the three antennas at 60 GHz by simulation. (a) Open-ended SIW, (b) ME dipole without vertical walls, and (c) ME dipole with vertical walls [16]

and fluctuates in gain over the operating band. By adding the two electric dipoles to form a complementary antenna structure, the antenna exhibits 46.5% impedance bandwidth and a stable gain up to 7.3 dBi with 2.3 dB variation over the working frequencies. Further improvement can be observed with the incorporation of the vertical walls.

Simulated radiation patterns of the three cases of antenna design are depicted in Figure 7.23. It is seen that the front-to-back ratio of the ME dipoles (Figure 7.23(b) and (c)) is substantially improved compared with the open-ended waveguide (Figure 7.23(a)). The ME dipoles have almost identical radiation patterns in the two principal planes with 80-degree beamwidth. This element is highly useful for the realization of multi-beam antenna arrays.

Other research groups have also investigated end-fire ME dipole antennas and arrays [17, 18]. The structures are more complicated to construct as the PCB layers for the feed network and the ME dipole element are perpendicular to each other.

7.4   ME Dipoles Fed by Gap Waveguides

The gap waveguide (GWG) was invented as an alternative transmission line with low transmission loss and high fabrication tolerance characteristics. In its basic form, it can be realized as a pure metallic transmission line. The GWG has been applied to excite ME dipole antennas and arrays.

In [19], an 8-element CP aperture-coupled ME dipole linear array excited by a GWG feed network was investigated. The antenna array exhibits 18 dBi gain in the boresight and an AR bandwidth of 14.5% with AR < 3 dB from 89 to 103 GHz. The work was also extended to realize a two-dimensional ME dipole array with similar performance [20].

Another group from Chalmers University of Technology also applied the single-layer ridge gap waveguide (RGW) to fabricate feed networks for exciting ME dipole antennas and arrays [21, 22]. In a recent design of an 8 × 8 CP aperture-coupled ME dipole arrays operated at around 28 GHz [23], a peak radiation efficiency and aperture efficiency of about 93% and 87%, respectively, were demonstrated. The gain is more than 26 dBic over a 20% bandwidth. This antenna array is suitable for Ka-band satellite communications.

For low-cost fabrication, the printed RGW was used to develop ME dipole arrays by a research group in Concordia University under the leadership of Prof. A. A. Kishk and Prof. A. R. Sebak. A typical structure is depicted in Figure 7.24. In [24], a GWG-fed ME dipole array with a metalens consisting of three layers of split ring resonators for gain enhancement was presented. The array exhibits SWR < 2 from 26.5 to 38.3 GHz. The antenna gain is more than 15 dBi between 28 and 38 GHz. The highest radiation efficiency is about 90%, occurred at 30 GHz. The work was also extended to realize a dual-polarized design with similar performance in gain and bandwidth [25]. Through an efficient procedure to design a large antenna array, a 16 × 16 ME dipole array fed by a printed RGW network was designed with 30 dBi

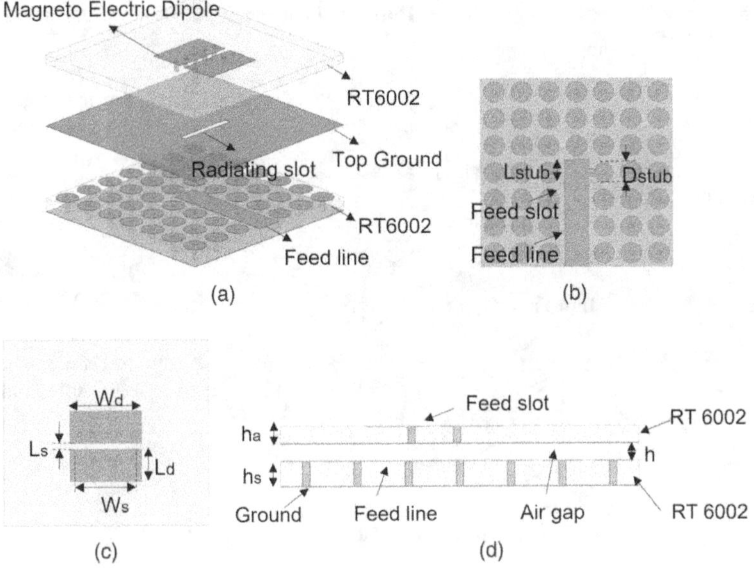

Figure 7.24 Typical ME dipole fed by a printed ridge gap waveguide. (a) Perspective view, (b) Ridge gap waveguide, (c) ME dipole, and (d) Side view [27].

gain and 71% radiation efficiency operating from 56 to 66 GHz [26]. To achieve 2D scanning, a 2 × 2 ME dipole antenna array fed by a printed RGW Butler Matrix operated at about 30 GHz was investigated. About 20% impedance bandwidth and stable gain of about 10.3 dBi were achieved [27].

Aperture-coupled ME dipole is usually not very wide in bandwidth. To achieve wider bandwidth, a modified gap waveguide was developed to achieve 28.8% impedance bandwidth with SWR < 2 from 22.6 to 30.2 GHz by Prof. Dongya Shen's group [28].

7.5 ME Dipole Excited by Other Kinds of Transmission Lines

To achieve wideband performance, both the antenna elements and feed networks are required to be wide in bandwidth. The SIW and GWG are intrinsically limited in bandwidth. The microstrip line feed network can be used to excite ME dipole antennas and arrays with very wideband performance, but the antenna cannot be placed on a mounting body too closely [29, 30]. Instead, the packaged microstrip line has been successfully employed to realize wideband ME dipole arrays. As shown in Figure 7.25, the basic element consists of four PCB layers. The top layer is for supporting an L-probe fed ME dipole. The second layer from the top is for supporting the microstrip line. The third layer from the top is a pure dielectric substrate for separating the second layer and the bottom layer, which is for realizing a periodic electromagnetic bandgap (EBG) layer. The EBG layer

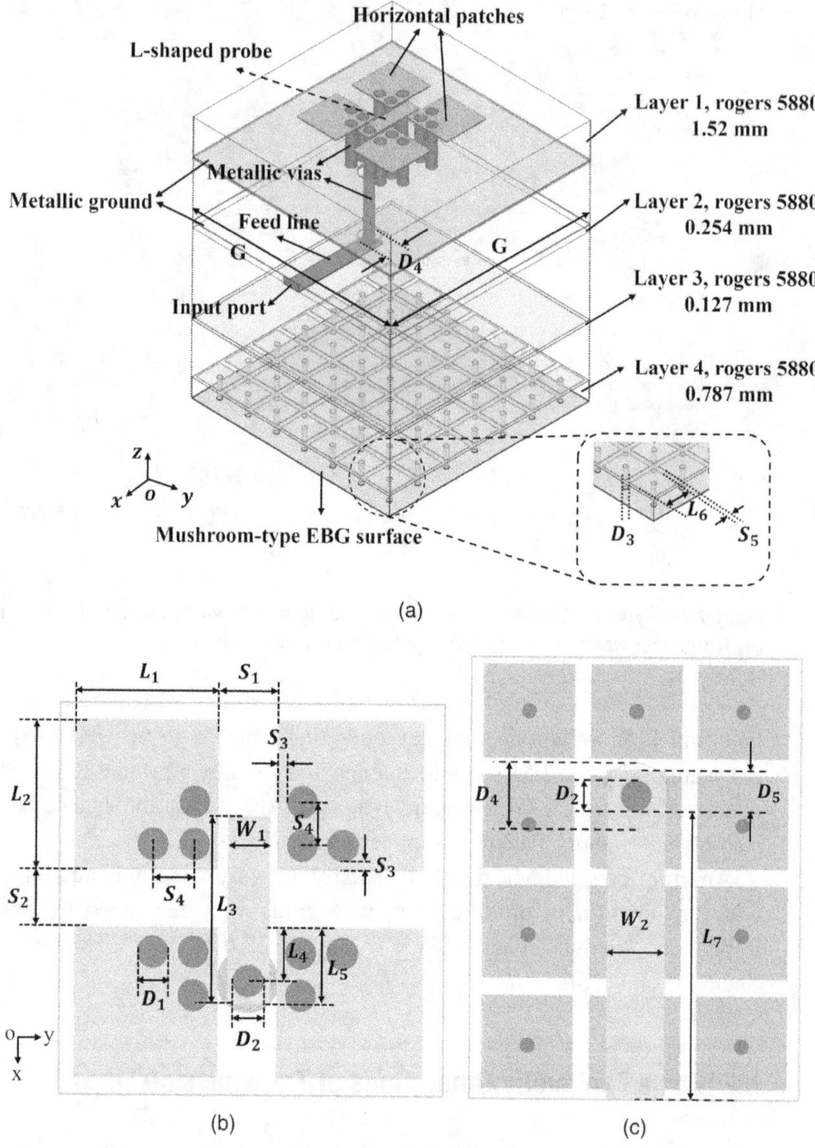

Figure 7.25 ME dipole fed by packaged microstrip line feed. (a) Perspective view, (b) Top view, and (c) Microstrip line [31].

functions as an artificial magnetic conductor (AMC). As shown in the dispersion diagram, Figure 7.26, EM waves cannot be propagated between the AMC and the grounded plane over a wide frequency range. A 4 × 4 ME dipole array, as shown in Figure 7.27, was designed and the performance was highly encouraging. With appropriate dimensions for the antenna structure, a wide impedance bandwidth of 51.5% with SWR < 2 from 24.5 to 40.5 GHz, a peak gain of 20.3 dBi, and a radiation efficiency of 85% at 32 GHz were achieved. A dual-polarized aperture-coupled

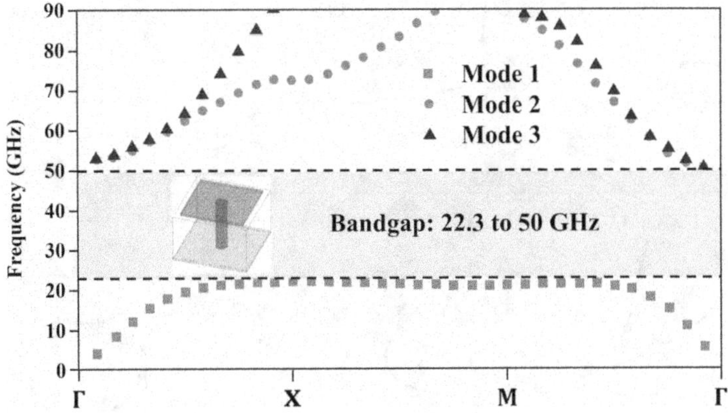

Figure 7.26 Dispersion diagram of EBG structure [31]

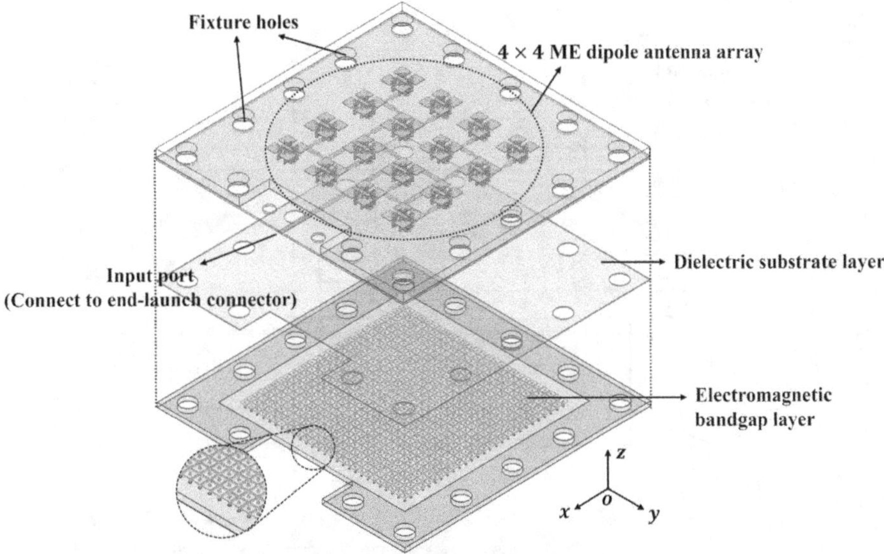

Figure 7.27 ME dipole array fed by packaged microstrip line network [31]

ME dipole array fed by a single layer of packaged microstrip line feed network was also developed with attractive performance [32].

Another approach for exciting ME dipoles with wideband performance is to use the SICL. With appropriate dimensions, the compact aperture-coupled ME dipole with SICL feed, as shown in Figure 7.28, has an impedance bandwidth of 48.8% with SWR < 2 from 24.3 to 40 GHz [33]. The antenna element is a good candidate for beam-steerable antenna array designs with wide scan angle and wideband performance.

End-fire ME dipole arrays fed by asymmetrical SICL [34] and symmetrical SICL [35] with wide bandwidth performance were also reported. The result is of

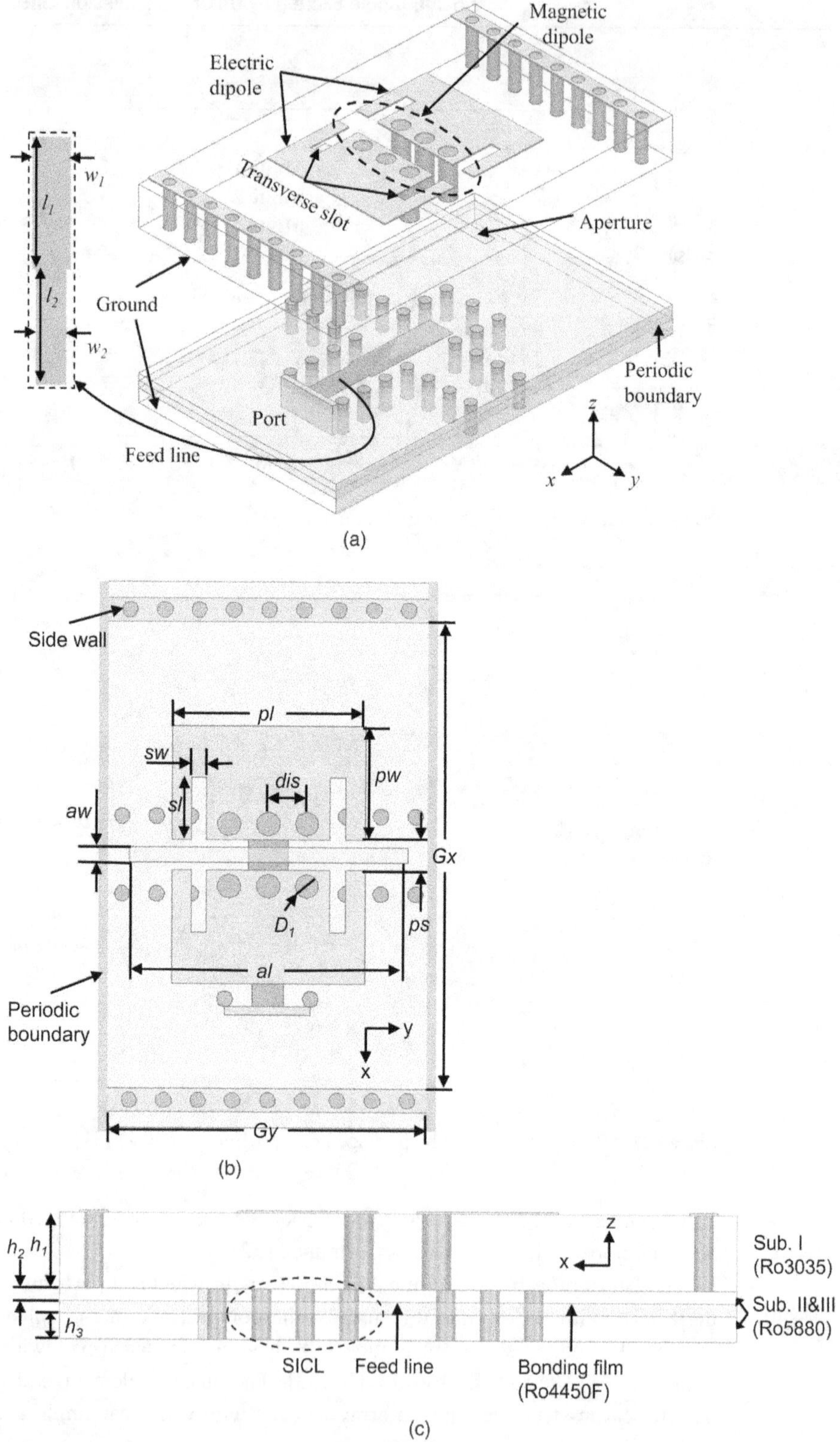

Figure 7.28 A compact ME-dipole antenna with SICL feed. (a) Perspective view, (b) Top view, and (c) Side view. (h_1 = 0.76 mm, h_2 = 0.127 mm, and h_3 = 0.254 mm) [33]

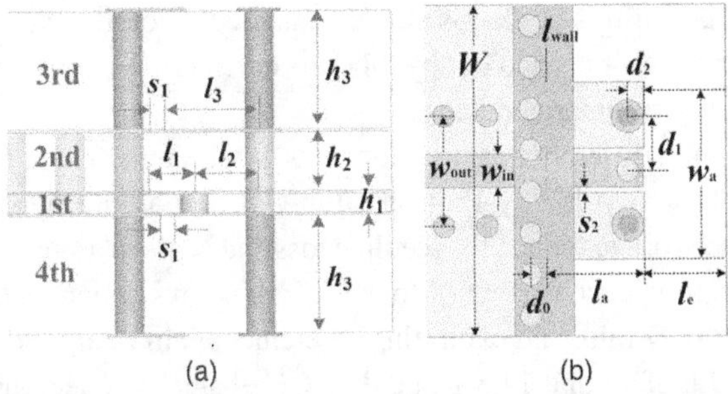

Figure 7.29 Geometry of end-fire ME dipole excited by asymmetrical coaxial line. (a) Side view and (b) Top view [34]

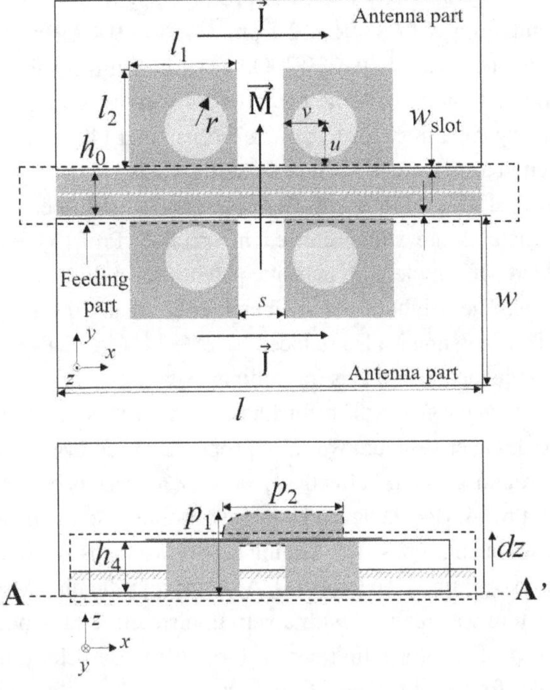

Figure 7.30 Geometry of an end-fire ME dipole excited by SICL [35]

practical importance (Figure 7.30). Magnetoelectric dipole excited by coplanar waveguide feed was also investigated [36]. Slightly wider bandwidth was achieved, but the antenna cannot be placed directly on a mounting structure.

7.6   Other Recent Interesting Works

The research and development of millimeter-wave ME dipole antennas and arrays is reviewed. The presentation is categorized by different kinds of feedlines used. Works conducted by the author of this book, his students, and collaborators are highlighted. Many more interesting works carried out by other research groups can be found in the literature and are briefly summarized later.

Methods of combining different transmission lines to form an efficient feed network were investigated. The group led by Prof. Tie Jun Cui proposed a hybrid SIW and microstrip line feed network to reduce the feedline loss and achieve a symmetrical feeding distribution. A prototype, consisting of four 2×2 subarrays, having a package size of 20 mm $\times$ 20 mm $\times$ 0.97 mm, and exhibiting an element gain of around 6.2 dBi and a percentage bandwidth of around 14.5% at the 60 GHz band, was demonstrated [37]. The group led by Prof. Jiro Hirokawa developed a wideband ME dipole array based on the combination of the ridge and groove GWGs operating at the Q-band. This antenna array is a full-metal design with a simple structure for achieving high efficiency and low-cost fabrication at very high frequencies. A prototype of an 8×8 ME dipole array exhibiting 32.5% bandwidth with SWR < 2 from 34.5 to 49.4 GHz, 27.5 dBi gain, and 75% antenna efficiency was shown [38]. Other earlier full-metal designs having the advantages of high gain and high efficiency for 5G wireless communications [39] and high-power handling capability for K-band airborne radar [40] can be found in the literature. These antenna arrays are 3D printable at low costs.

Although the physical size of ME dipoles operating at millimeter-wave is very small, low-profile structures are still preferred in practice. Prof. Yue Ping Zhang's group designed a folded ME dipole with parasitic patches for millimeter-wave antenna-in-package applications. The height of the antenna is reduced to 10% free-space wavelength and the size of the antenna is also reduced by 25% [41]. A dual-polarized folded ME dipole with two-stage meandered vias for millimeter-wave 5G applications was proposed by Jenn-Hwan Tarng's group. The thickness is reduced to 11% wavelength in dielectric [42]. A double-layer structure was also proposed to achieve a 1 mm thickness design operating at around 26 GHz, effectively about 8% free-space wavelength, for 5G smartphones [43]. Prof. I. Boccia developed a low-profile design with a thickness of about 9% free-space wavelength based on adding capacitive loads between the electric dipole arms and defective ground plane [44, 45]. A frequency-scanning ME dipole array based on a comb structure was realized with a very thin structure of about 6.5% wavelength thickness by Prof. Ronghong Jin's group [46]. All these achievements help to promote the ME dipoles for the mmWave 5G and future 6G wireless communications.

Antennas with filtering function are highly attractive in system design as it reduces the stringent requirement on the filters connecting to the antennas. Promising designs of filtering ME dipoles were reported by Prof. Wolfgang Bosch's group [47], Prof. Xiuyin Zhang's group [48], and Prof. Aarno Parssinen's group [49].

More recently, the millimeter-wave ME dipoles have been applied for advancing the designs of reflectarrays and transmitarrays [50, 51], Yagi antennas [52], AiP end-fire arrays for 5G user equipment [53], shared-aperture antennas [54], dual-circularly polarized antenna arrays [55], frequency scanning antenna arrays [56], high-gain interconnected antenna arrays [57, 58], non-Foster antennas [59], low-cost high-gain circularly polarized antenna arrays [60], antennas for energy harvesting and wireless power transfer [61, 62], wide beamwidth wideband antennas [63, 64], wide scan antenna arrays for 5G wireless devices [65, 66, 67], high-gain antenna for 5G back-haul systems [68], antennas for K/Ka-band satellite communications [69], metasurface antennas [70], and terahertz antennas operating at F-band [71], and D-band [72, 73]. With concerted efforts by academic scholars and practicing engineers across the globe, more robust and efficient ME dipole antennas and arrays will be developed in the foreseeable future.

7.7   Summary

The research and development of millimeter-wave ME dipole antennas and arrays is reviewed. The presentation is categorized by different kinds of feedlines used. Works conducted by the author of this book, his students, and collaborators are highlighted. Many other recent interesting works carried out by other research groups found in the literature are briefly summarized.

References

[1] K. B. Ng, H. Wong, K. K. So, C. H. Chan and K. M. Luk, 60 GHz plated through hole printed magneto-electric dipole antenna, *IEEE Transactions on Antennas and Propagation*, vol. 60, 2012, no. 7, pp. 3129–3136.

[2] M. Li and K. M. Luk, Wideband magneto-electric dipole antenna for 60-GHz millimeter-wave communications, *IEEE Transactions on Antennas and Propagation*, vol. 63, 2015, no. 7, pp. 3276–3279.

[3] M. Li and K. M. Luk, A wideband circularly polarized antenna for microwave and millimeter-wave applications, *IEEE Transactions on Antennas and Propagation*, vol. 62, 2014, no. 4, pp. 1872–1879.

[4] J. Wu, W. Wu and J. Li, Dual-polarized magneto-electric dipole antenna and MIMO array for 5G millimeter-wave applications, *2019 International Conference on Microwave and Millimeter Wave Technology*, Guangzhou, China, May 2019, pp. 1–3.

[5] Y. Li, C. Wang and Y. X. Guo, A Ka-band wideband dual-polarized magneto-electric dipole antenna array on LTCC, *IEEE Transactions on Antennas and Propagation*, vol. 68, 2020, no. 6, pp. 4985–4990.

[6] S. Chen and A. Zhao, LTCC based dual-polarized magneto-electric dipole antenna for 5G millimeter wave application, *13th European Conference on Antennas and Propagation*, Krakow, Poland, 2019, pp. 1–4.

[7] Y. Wang, M. Li, H. Liu, Y. Shi and L. Li, Wideband substrate-integrated-waveguide-fed magneto-electric dipole array antenna, *2019 International Conference on Microwave and Millimeter Wave Technology*, Guangzhou, May 2019, pp. 1–3.

[8] Z. C. Hao, Q. Yuan, B. W. Li and G. Q. Luo, Wideband W-band substrate-integrated waveguide magnetoelectric (ME) dipole array antenna, *IEEE Transactions on Antennas and Propagation*, vol. 66, 2018, no. 6, pp. 3195–3200.

[9] Q. Zhu, K. B. Ng, C. H. Chan and K. M. Luk, Substrate-integrated-waveguide-fed array antenna covering 57–71 GHz band for 5G applications, *IEEE Transactions on Antennas and Propagation*, vol. 65, 2017, no. 12, pp. 6298–6306.

[10] Y. Li and K. M. Luk, A 60-GHz wideband circularly polarized aperture-coupled magneto-electric dipole antenna array, *IEEE Transactions on Antennas and Propagation*, vol. 64, 2016, no. 4, pp. 1325–1327.

[11] Y. F. Wang, B. Wu, N. Zhang, Y. T. Zhao and T. Su, Wideband circularly polarized magneto-electric dipole 1 × 2 antenna array for millimeter-wave applications, *IEEE Access*, vol. 8, 2020, pp. 27516–27523.

[12] B. Feng, J. Lai, K. L. Chung, T. Y. Chen, Y. Liu and C.-Y.-D. Sim, A compact wideband circularly polarized magneto-electric dipole antenna array for 5G millimeter-wave applications, *IEEE Transactions on Antennas and Propagation*, vol. 68, 2020, no. 9, pp. 6838–6843.

[13] Y. Li and K. M. Luk, 60-GHz dual-polarized two-dimensional switch-beam wideband antenna array of aperture-coupled magneto-electric dipoles, *IEEE Transactions on Antennas and Propagation*, vol. 64, 2016, no. 2, pp. 554–563.

[14] Y. Li, J. Wang and K. M. Luk, Millimeter-wave multi-beam aperture-coupled magneto-electric dipole array with planar substrate integrated beamforming network for 5G applications, *IEEE Transactions on Antennas and Propagation*, vol. 65, 2017, no. 12, pp. 6422–6431.

[15] Y. Li and K. M. Luk, A multibeam end-fire magnetoelectric dipole antenna array for millimeter-wave applications, *IEEE Transactions on Antennas and Propagation*, vol. 64, 2016, no. 7, pp. 2894–2904.

[16] J. Wang, Y. Li, L. Ge, J. Wang and K. M. Luk, A 60 GHz horizontally polarized magneto-electric dipole antenna array with 2-D multibeam endfire radiation, *IEEE Transactions on Antennas and Propagation*, vol. 65, 2017, no. 11, pp. 5837–5845.

[17] W. Jiang, Y. Cui and T Zhang, Eight-element linear magneto-electric dipole antenna array with Butler matrix feed network for 5G applications, *2019 International Symposium on Antennas and Propagation*, Oct 2019.

[18] Q. F. Fu, Z. H. Tu and Z. Gan, Millimeter-wave end-fire circular-polarized ME dipole antenna, *2019 International Conference on Microwave and Millimeter Wave Technology*, May 2019.

[19] J. Cao, H. Wang, S. Mou, S. Quan and Z. Ye, W-band high-gain circularly polarized aperture-couple magneto-electric dipole antenna array with gap waveguide feed network, *IEEE Antennas and Propagation Letters*, vol. 16, 2017, pp. 2155–2158.

[20] J. Cao, H. Wang, S. Mou, P. Soothar and J. Zhou, An air cavity-fed circularly polarized magneto-electric dipole antenna array with gap waveguide technology for mm-wave applications, *IEEE Transactions on Antennas and Propagation*, vol. 67, 2019, pp. 6211–6216.

[21] W. Y. Yong, T. Emanuelsson and A. A. Glazunov, 4 × 4 magneto-electric dipole array with single-layer corporate-feed ridge gap waveguide for mmWave applications, *2020 International Symposium on Antennas and Propagation*, 2020, pp. 573–574.

[22] W. Y. Yong, T. Emanuelsson and A. A. Glazunov, 5G wideband magneto-electric dipole antenna fed by a single-layer corporate-feed network based on ridge gap waveguide, *2020 European Conference on Antennas and Propagation*, 2020, pp. 1–4.

[23] L. Sun, A. Uz Zaman, B. Yan and W. Sheng, Wideband high-efficiency single-layer ridge gap waveguide based circular polarized magneto-electric dipole antenna array, *IEEE Transactions on Antennas and Propagation*, vol. 72, 2024, no. 8, pp. 6735–6740.

[24] A. Dadgarpour, M. S. Sorkherizi and A. A. Kishk, Wideband low-loss magnetoelectric dipole antenna for 5G wireless network with gain enhancement using meta lens and

gap waveguide technology feeding, *IEEE Transactions on Antennas and Propagation*, vol. 64, 2016, no. 12, pp. 5094–5101.

[25] A. Dadgarpour, N. Bayat-Makou, M. A. Antoniades, A. A. Kishk and A. Sebak, A dual-polarized magneto-electric dipole array based on printed ridge gap waveguide with dual-polarized split-ring resonator lens, *IEEE Transactions on Antennas and Propagation*, vol. 68, 2020, no. 5, pp. 3578–3585.

[26] A. T. Hassan and A. A. Kishk, Efficient procedure to design large finite array and its feeding network with examples of ME-dipole array and microstrip ridge gap waveguide feed, *IEEE Transactions on Antennas and Propagation*, vol. 68, 2020, no. 6, pp. 4560–4570.

[27] M. M. M. Ali and A. R. Sebak, 2-D scanning magnetoelectric dipole antenna array fed by RGW Butler matrix, *IEEE Transactions on Antennas and Propagation*, vol. 66, 2018, no. 11, pp. 6313–6321.

[28] C. Ma, M. Dong and D. Shen, A printed magneto-electric dipole based on substrate integrated gap waveguide, *2017 Sixth Asia-Pacific Conference on Antennas and Propagation*, Xian, Oct 2017.

[29] C. Zhang, R. Zhang, J. Zheng, Y. Shao and F. Lin, A millimeter-wave windmill-shaped magnetoelectric-dipole antenna with broadband circular polarization and wide axial-ratio beamwidth, *2019 International Conference on Computer Intelligent Systems and Network Remote Control*, 2019, pp. 549–554.

[30] D. T. T. My, H. N. B. Phuong, T. T. Huong and B. T. M. Tu, A magneto-electric dipole antenna array for millimeter wave applications, *Engineering, Technology & Applied Science Research*, vol. 10, 2020, no. 4, pp. 6057–6061.

[31] J. Sun, A. Li and K. M. Luk, A high-gain millimeter-wave magnetoelectric dipole array with packaged microstrip line feed network, *IEEE Antennas and Wireless Propagation Letters*, vol. 19, 2020, no. 10, pp. 1669–1673.

[32] N. Ashraf and A. R. Sebak, Packaged microstrip line feed network on a single surface for dual-polarized $2^N \times 2^M$ ME-dipole antenna array, *IEEE Antennas and Wireless Propagation Letters*, vol. 19, 2020, no. 4, pp. 596–600.

[33] X. Dai, A. Li and K. M. Luk, A wideband compact magnetoelectric dipole antenna fed by SICL for millimeter wave applications, *IEEE Transactions on Antennas and Propagation*, vol. 69, 2021, no. 9, pp. 5278–5285.

[34] A. Li and K. M. Luk, Millimeter-wave end-fire magneto-electric dipole antenna and arrays with asymmetrical substrate integrated coaxial line feed, IEEE, *Open Journal of Antennas and Propagation*, vol. 2, 2021, pp. 62–71.

[35] J. Yin, Q. Wu, C. Yu, H. Wang and W. Hong, Broadband endfire magnetoelectric dipole antenna array using SICL feeding network for 5G millimeter-wave applications, *IEEE Transactions on Antennas and Propagation*, vol. 67, 2019, no. 7, pp. 4895–4900.

[36] Z. C. Hao and B. W. Li, Developing wideband planar millimeter-wave array antenna using compact magneto-electric dipoles, *IEEE Antennas and Wireless Propagation Letters*, vol. 16, 2017, pp. 2102–2105.

[37] T. Zhang, Z. Zhu, H. Xia, W. Xu, L. Li and T. J. Cui, 60-GHz scalable LTCC phased array with compact symmetric hybrid feeding network for antenna-in-package application, *IEEE Transactions on Components Packaging and Manufacturing Technology*, vol. 13, 2023, no. 13, pp. 1694–1702.

[38] Y. Wu, T. Tomura, J. Hirokawa and M. Zhang, A wideband full-metal sidewall-loaded magnetoelectric dipole array based on combined ridge and groove gap waveguides in the Q-band, *IEEE Transactions on Antennas and Propagation*, vol. 71, 2023, no. 7, pp. 6156–6161.

[39] K. K. So, K. M. Luk, C. H. Chan and K. F. Chan, 3D printed high gain complementary dipole/slot antenna array, *Applied Sciences*, vol. 8, 2018, p. 1410.

[40] B. Zhang, H. Sun, L. Wu, Y. Zhou, Y. Yang, H. Zhu, F. Cheng, Y. He and K. Huang, A metallic 3-D printed airborne high-power handling magneto-electric dipole array with cooling channels, *IEEE Transactions on Antennas and Propagation*, vol. 67, 2019, no. 12, pp. 7368–7378.

[41] W. Wang, M. Tang, Z. Shao and Y. Zhang, A low-profile magneto-electric dipole antenna with parasitic patches for millimeter-wave antenna-in-package applications, *2019 Photonics & Electromagnetics Research Symposium*, Xiamen, Dec 2019, pp. 3016–3019.

[42] Y. C. Chang, C. C. Hsu, M. Idrees Magray, H. Y. Chang and J. H. Tarng, A novel dual-polarized wideband and miniaturized low profile magneto-electric dipole antenna array for mmWave 5G applications, *IEEE Open Journal of Antennas and Propagation*, vol. 2, 2021, pp. 326–334.

[43] Z. W. Yin, L. Wang, W. W. Yang and J. Shi, Magneto-electric dipole 26 GHz phased array for 5G application of smartphones, *2018 11th UK-Europe-China Workshop on Millimeter Waves and Terahertz Technologies*, Hang Zhou, Sept 2018.

[44] G. Scalise, L. Boccia and G. Amendola, An ultralow profile magneto electric dipole for 5G applications, 2019 *14th International Conference on Advanced Technologies, Systems and Services in Telecommunications (TELSIKS)*, Nis, Serbia, 2019, pp. 141–143.

[45] G. Scalise, L. Boccia, G. Amendola, M. Rousstia, and A. Shamsafar, Magneto-electric dipole antenna for 5-G applications, *2020 14th European Conference on Antennas and Propagation*, Copenhagen, Mar 2020, pp. 1–3.

[46] R. Zhao, J. Geng, H. Zhou, S. Yang, J. Lu, X. Tang, N. Chen, A. Zhang, H. Li, X. Li, E. Li, C. He and R. Jin, A low-profile magneto-electric dipole antenna based on comb structure and its application in frequency -scanning array with ground, *IEEE Antennas and Wireless Propagation Letters*, vol. 23, 2024, no. 10, pp. 2855–2859.

[47] H. S. Farahani, B. Rezaee and W. Bosch, Ka-band coupled-resonator filtering magneto-electric dipole antenna, *Proceedings of the 50th European Microwave Conference*, 2020, pp. 722–725.

[48] S. Wu, J. Li, X. Chen, S. Yan and X. Y. Zhang, A Ka-band SLM printed filtering divider-fed magnetoelectric dipole antenna array using embedded gap waveguide, *IEEE Antennas and Wireless Propagation Letters*, vol. 22, 2023, no. 4, pp. 774–778.

[49] J. Chen, M. Berg, K. Rasilainen, Z. Siddiqui, M. E. Leinonen and A. Pärssinen, Integration of second-order Bandstop filter into a dual-polarized 5G millimeter-wave magneto-electric dipole antenna, *IEEE Transactions on Antennas and Propagation*, vol. 72, 2024, no. 6, pp. 5361–5366.

[50] K. M. Luk and B. Xiang, Transmitarray and reflectarray antennas based on a magnetoelectric dipole antenna, Electromagnetic, *Science*, vol. 1, 2023, no. 1, p. 0010091.

[51] S. Huang, Z. Wu, M. Wang, Y. Pu, Z. Wen, J. Xu, J. Wang and Y. Luo, Single-layer multimode broadband circularly polarized reflected element using characteristic mode analysis, *IEEE Antennas and Wireless Propagation Letters*, vol. 23, 2024, no. 9, pp. 2787–2791.

[52] Q. Q. Wang, Z. P. Qian, W. Q. Cao, Y. S. Zhang and K. Li, 60GHz stacked Yagi Magneto-Electric Dipole antenna with wideband and high gain properties, 2015 *IEEE International Conference on Communication Problem-Solving (ICCP)*, Guilin, China, 2015, pp. 454–457.

[53] H. T. Chou, S. J. Chou, J. D. S. Deng, C. H. Chang and Z. D. Yan, LTCC-based antenna-in-package array for 5G user equipment with dual-polarized endfire radiations at millimeter-wave frequencies, *IEEE Transactions on Antennas and Propagation*, vol. 70, 2022, no. 4, pp. 3076–3081.

[54] Y. F. Cheng, J. R. Wu, L. Peng, C. Liao and X. Ding, A printed-RGW-fed shared-aperture antenna integrating Ku-band magneto-electric dipole and Ka-band Fabry-Perot radiator, *IEEE Antennas and Wireless Propagation Letters*, vol. 23, 2024, no. 9, pp. 2772–2776.

[55] W. Zeng, X. Y. Wu, F. Wu, Y. Zhang, Z. H. Jiang, W. Hong and K. M. Luk, Broadband dual-CP multi-stage sequential rotation arrays with independent control of polarizations based on dual-CP magneto-electric dipole elements, *IEEE Transactions on Antennas and Propagation*, vol. 72, 2024, no. 4, pp. 3017–3032.

[56] X. X. Yang, H. Qiu, T. Lou, Z. Yi, Q. D. Cao and S. Gao, Circularly polarized millimeter wave frequency beam scanning antenna based on aperture-coupled magneto-electric dipole, *IEEE Transactions on Antennas and Propagation*, vol. 70, 2022, no. 9, pp. 7603–7611.

[57] Z. Chen, W. Zhang and K. X. Wang, A multi-port interconnected magneto-electric dipole antenna array for 5G applications, *Electronics Letters*, vol. 59, 2023, no. 5, pp. 1–3.

[58] S. J. Yun, K. H. Lee, J. N. Lee and Y. K. Cho, An inclined bridge-connected magneto-electric dipole array for millimeter-wave broadband applications, *IEEE Antennas and Propagation Letters*, vol. 23, 2024, no. 7, pp. 2140–2144.

[59] D. Nozina, J. Bartolic and S. Hrabar, Active non-Foster-based magneto-electric antenna, *17th European Conference on Antennas and Propagation*, Florence, Italy, 2023.

[60] Y. Al-Alem, S. M. Sifat, Y. M. M. Antar and A. A. Kishk, Circularly polarized Ka-band high-gain antenna using printed ridge gap waveguide and 3D printing technology, *IEEE Transactions on Antennas and Propagation*, vol. 71, 2023, no. 9, pp. 7644–7649.

[61] F. Deng and K. M. Luk, A broadband high-gain multi-beam ambient millimeter-wave energy harvesting system, *IEEE Internet of Things Journal*, vol. 11, 2024, no. 3, pp. 4888–4898.

[62] J. Sang, L. Qian, M. Li, J. Wang and Z. Zhu, A wideband and high-gain circularly polarized antenna array for radio-frequency energy harvesting applications, *IEEE Transactions on Antennas and Propagation*, vol. 71, 2023, no. 6, pp. 4874–4887.

[63] Y. Zhao, Z. Bao and Y. Li, 77 GHz Ultrawide-beam magneto-electric dipole antenna array with high isolation, *2022 IEEE MTT-S International Microwave Workshop Series on Advanced Materials and Processes for RF and THz Applications (IMWS-AMP)*, Guangzhou, China, 2022, pp. 1–3.

[64] P. Boontamchauy, M. Sano, R. Kuse and T. Fukusako, Circularly polarized cavity-backed antenna with variable magneto-electric crossed-dipole structure, *IEICE Communications Express*, vol. 12, 2024, pp. 1–4.

[65] M. Doring, N. Kastle, T. Frey, F. Matt, C. Waldschmidt and T. Chaloun, Low-profile wide-scan Magneto-Electric Dipole antenna for 5G mm-wave communications, *18th European Conference on Antennas and Propagation (EuCAP)*, Apr 2024.

[66] G. Scalise, E. Arnieri, G. Amendola, M. W. Rousstia, S. Pires and L. Boccia, Dual-polarized magneto-electric dipole for 5G 28-GHz phased array applications, *IEEE Open Journal of Antennas and Propagation*, vol. 5, 2024, no. 1, pp. 164–179.

[67] W. Y. Yong and A. A. Glazunov, Wideband 2 × 2 antenna-in-package based on magneto-electric dipole array antenna for 5G mmWave applications, *Frontiers in Antennas and Propagation*, vol. 2, 2024, pp. 1–17.

[68] C. Mustacchio, L. Boccia, E. Arnieri and G. Amendola, E Band high gain antenna for 5G Backhauling Systems, *18th European Conference on Antennas and Propagation*, Apr 2024.

[69] J. Wei, S. Liao, Q. Xue and W. Che, Wideband circularly polarized phased array antenna for K/Ka-Band satellite communication using ME-dipole elements, *IEEE Antennas and Wireless Propagation Letters*, vol. 23, 2024, no. 8, pp. 2496–2500.

[70] X. Ruan, K. Wang and Q. Zhang, Implementation of magnetoelectric dipoles in meta-surfaces for 5G applications, IEEE, *Open Journal of Antennas and Propagation*, vol. 3, 2022, pp. 1253–1263.

[71] F. Matt, D. Zimmermann, M. Döring and C. Waldschmidt, Broadband waveguide magneto-electric dipole antenna for F-band applications, 2024, *18th European Conference on Antennas and Propagation (EuCAP)*, Glasgow, United Kingdom, 2024, pp. 1–4.

[72] H. C. Kuo, C. W. Kuo, C. C. Wang and C. P. Hung, A D-band magnetoelectric dipole antenna-in-package (AiP) implemented on BT-based organic substrate, *IEEE Transactions on Components, packaging and manufacturing Technology*, vol. 12, 2022, no. 10, pp. 1743–1680.

[73] X. Zhong, Q. Li, C. Guo, J. Li, Z. Wang, J. Shi, X. Chen and A. Zhang, A D-band wideband magnetoelectric dipole antenna array based on micro-metal additive manufacturing, *IEEE Transactions on Antennas and Propagation*, vol. 72, 2024, no. 8, pp. 6500–6509.

8 Concluding Remarks

8.1 Summary of All Chapters

Most of the classical complementary antennas are narrow in bandwidth. This is due to the fact that narrow slots were used to realize magnetic currents. The ME dipole was proposed to fulfill the demand of wideband unidirectional antennas for base station of 3G mobile communication systems operating at microwave frequencies below a few GHz. The basic structure of the antenna looks like a reflector-backed planar electric dipole excited by an L-shaped probe-fed shorted patch antenna, which also radiates as a magnetic dipole together with the electric dipole. The ME dipole not only exhibits wide bandwidth but also has low cross polarization, low backside radiation, symmetrical radiation pattern, and stable gain and beamwidth over the operating frequencies. These attractive characteristics can be explained by the complementary nature of the antenna structure as an electric dipole mode and a magnetic dipole mode are excited together over the frequency band of operation. The approach can be considered as a generalization of the classical complementary antennas concept, developed in early 1970s, in which the electric and magnetic dipoles are excited at the same resonant frequency. In the ME dipole, the resonant frequency of the electric dipole mode is slightly lower than that of the magnetic dipole mode. This provides the amazing performance of stable gain and radiation pattern over a wide frequency range.

The ME dipole technology is versatile in applications, as many high-performance linearly polarized, circularly polarized, dual-polarized, low-profile, ultra-wideband, high-gain ME dipole elements with single feed or differential feeds were developed successfully over the years. In this book, the design and performance of the original ME dipole antenna have been reviewed. More detailed information is provided, including an improved equivalent circuit and current density distribution in various parts of the antenna. The development of the linearly polarized ME dipoles is provided. Techniques for single-band and dual-band operation are summarized. Available feeding structures, including the differential feeds, are studied. High gain antenna arrays of linearly polarized magnetoelectric dipoles are also reported. Several representatives of dual-polarized and circularly polarized ME dipoles are reviewed. Major techniques for reducing the height and projection area of the ME dipole are described. Most of the available and effective designs for ME dipoles with wide bandwidth or ultra-wide bandwidth are included. The design principle of each antenna is explained.

The advancement of millimeter wave and Terahertz technology to support the growth of 5G and beyond requires sophisticated high-gain antenna arrays with steerable or reconfigurable characteristics. The ME dipole is demonstrated to be a wonderful choice, as most of the ME dipole designs operating at lower microwave frequencies can be modified and enhanced for millimeter wave operation and can be realized by advanced fabrication processes, including low-cost 3D-printing techniques. The development of millimeter-wave ME dipole arrays with different kinds of feeding structures and transmission lines for diverse applications is reviewed.

8.2   Possible Future Directions

It is expected that ME dipoles will be applied continuously for diverse applications including 5G and 6G mobile communications, global positioning navigation systems, satellite internets, radars, medical imaging, and wireless power transfer and energy harvesting. It is also employed to advance novel antenna technology, including reconfigurable intelligent surfaces, metasurface antennas, transmitarrays and reflectarrays, antenna arrays with wide scanning angles, connected antenna arrays, different kinds of radars, and many more to come.

The success of the ME dipole motivated a renewed interest in applying the complementary antenna concept to develop more novel antennas for wireless communications and sensing, such as the planar ME dipoles, Huygens source antennas, and shared aperture antennas. It is anticipated that more novel antennas will be proposed based on the excitation of multiple electric and magnetic dipole modes to achieve special performance for various applications.

Index

For EU product safety concerns, contact us at Calle de José Abascal, 56–1°,
28003 Madrid, Spain or eugpsr@cambridge.org.